Genetic Algorithms in Applications

Genetic Algorithms in Applications

Editor

Asa Bensten

scitus academics

Genetic Algorithms in Applications
Edited by **Asa Bensten**

ISBN: 978-1-68117-263-7
Library of Congress Control Number: 2016937431

SCITUS Academics LLC,
www.scitusacademics.com
Box No. 4766, 616 Corporate Way,
Suite 2, Valley Cottage,
NY 10989

Printed in United States of America on Acid Free Paper

Preface

Genetic Algorithm (GA) is a process of artificial intelligence. It is based on theories of natural selection and development. It is a heuristic search technique used in computing and artificial intelligence to find optimized solutions to search problems using techniques inspired by evolutionary biology: mutation, selection, reproduction and recombination. Genetic algorithms have been used in science, engineering, business and social sciences. Traditional search and optimization methods are too slow to find a solution in a very complex search area even on supercomputers. Genetic algorithm is a robust search method that requires little information to search effectively in a large or incomprehensible search area. In particular, a genetic search progresses through a number of points as opposed to the single focus of the search algorithms. It is also useful in the non-linear problem area. Its intrinsic parallelism in evaluation and selection allows the use of distributed processing machines.

This book can be of immense use to engineers and scientists in different fields of expertise who need certain optimization techniques in their work and who are able to use Genetic Algorithms for the first time. These applications can be of helpful for many individuals who know the genetic algorithm already.

Table of Contents

1

CHAPTER

A NEW APPROACH TO THE OPTIMIZATION OF THE CVRP THROUGH GENETIC ALGORITHMS

Mariano Frutos[1], Fernando Tohmé[2]

Department of Engineering, National University of the South, Bahía Blanca, Argentina

[2]Department of Economics, National University of the South, Bahía Blanca, Argentina

ABSTRACT

This paper presents a new approach to the analysis of complex distribution problems under capacity constraints. These problems are known in the literature as CVRPs (Capacitated Vehicle Routing Problems). The procedure introduced in this paper optimizes a transformed variant of a CVRP. It starts generating feasible clusters and codifies their ordering. In the next stage the procedure feeds this information into a genetic algorithm for its optimization. This makes the algorithm independent of the constraints and improves its performance. Van Breedam problems have been used to test this technique. While the results obtained are similar to those in other works, the processing times are longer.

Keywords: Vehicle Routing Problem; Genetic Algorithms; Modeling; Optimization

1. INTRODUCTION

The complexity of distribution problems is the reason for the lack of solution techniques able to provide optimal solutions to real cases in a reasonable time [1]. But firms have to make decisions on transportation matters, sometimes in the short run. The economic importance of moving people and goods makes it necessary the development of strategies to generate quick competitive solutions [2,3]. In the literature, these problems are generically known as the Vehicle Routing Problem (VRP) [4], which is NP-Complete combinatorial optimization problem. Its scholarly and practical interest stems from the aforementioned lack of quick solution methods [5]. The main goal of this paper is to introduce a new algorithm, robust, flexible and fast enough to solve different instances of this problem.

The rest of the paper is organized as follows. Firstly, we present some characterizations of the VRP as well as of the original version of the CVRP (Capacitated Vehicle Routing Problem). Then, we introduce an alternative version of the latter and a methodology for its solution. We present later the result of the experiments with this method. Finally we discuss possible further work on the subject.

2. DEFINITION OF CVRP

In rough terms, the statement of the VRP is: given classes of clients and of storage sites (both scattered over a geographical region) and a fleet of vehicles, determine the minimal cost routes starting and ending in storage sites delivering the goods to the clients. The features of clients, storage sites and vehicles, as well as the operational constraints over routes determine the different variants of the problem. Each client is associated to a demand that has to be covered by a vehicle. This demand is usually understood as an amount of a good that occupies some volume of the vehicle. In crowded area a single vehicle may not be able to satisfy the demands of all the clients living there. The capacity of the vehicle imposes then a bound on the number of clients that it might satisfy at the same time [6,7]. This kind of constraints defines the CVRP (Capacitated Vehicle Routing Problem). The capacity of the vehicle might be in terms of weight or volume. When more than a single kind of products is transported the vehicles could be compartmented carrying each type of parcel in a different subdivision. A distinction can be made between the problems of homogeneous fleets, in which the attributes (capacity, cost, etc.) are the same for all the vehicles, and heterogeneous fleet problems. The latter case raises the possibility of incompatibilities between the capacities of the vehicles and the demands of the clients. If then, each client might be visited only by some of the vehicles. It could also happen that the goods to be transported might not be available in a single storage site and must be collected from several sites. All these depots have to be visited before delivering the goods to the client. While normally each client should be visited once, it could be necessary to satisfy his demand at different moments and by different vehicles. Some clients could have some hourly restrictions for the reception of parcels, which are expressed as time windows [8,9]. In such case the problem is known as VRPTW (Vehicle Routing Problem with Time Windows). The vehicles, as well as the products, are usually parked at the storage sites. It is also usually required that the vehicle starts and ends trips from the same storage site, but this restriction can be lifted in some cases. The number of available vehicles could be given or be decision variable [10]. The objective is frequently the minimization of the distances as well as the number of vehicles used while satisfying the demands of the clients [11,12]. Recent studies have allowed the possibility of vehicles traversing different routes [13]. Different presentations of such problems can be found in the literature [14-17], as well as different approaches to solving them [18-23].

In this work we focus on the CVRP. Consider version (1) of the problem. Let N^c be the class of client nodes, endowed with a linear order <, N^s the set of storage nodes while $N = N^c \cup N^s$. C_{ij} the cost (distance) of going from nodes i to j for any pair $(i,j) \in N \times N$. X_{ij} is a binary variable, which is 1 if a vehicle goes from i to j without intermediate stops. Otherwise $X_{ij} = 0$. $C_{\max}$ is the maximum capacity of a vehicle and d_i the amount demanded at a node $i \in N^c$. Let S_j be the class of nodes in a path that starts and ends at node $j \in N^s$. Then, the problem is:

$$\min \sum_{i \in N} \sum_{j \in N} C_{ij} X_{ij} \tag{1}$$

s.t.:

1) $\sum_{i \in N^c \cap S_j} d_i \leq C_{\max}$ 2) and $\{S_j\}_{j \in N^s}$ such that $(S_i \cap S_j) \cap N^c = \varnothing$

In words: minimize the total distance traversed by all vehicles, satisfying the demands on every route for a partition of client nodes in clusters, each one covered by a single vehicle. Notice that constraint 2) makes the problem more complex than a linear programming one.

Version (2) is as follows: let $M = \{S_j\}_{j \in N^s}$ for some allocation of routes, $c_m = \sum_{(i,j) \in m} C_{ij}$ the sum of distances over route m, X_m is a binary variable that is 1 if route m is used and 0 otherwise and e_{im} also a binary variable (1 if $i \in N^c$ belongs to route m and 0 otherwise). Then, the problem is:

$$\min \sum_{m \in M} c_m X_m \qquad (2)$$

s.t.:

3) and for each i, $\sum_{m \in M} e_{im} X_m = 1$ That is (2) is analogous to (1) under constraint 1) with constraint 2) already satisfied.

3. METHODOLOGY

An algorithm solving (2) proceeds in two stages. The first one generates feasible clusters, classifying and ordering them. The second stage solves the minimization problem by means of a Genetic Algorithm [24].

3.1. First Stage

The algorithm generates the clusters that will be elements of $\{S_j\}_{j \in N^s}$ .We take each $i \in N^c$ and consider all the clusters S_j such that $i \in S_j$. The inductive construction starts with i = 1 (the lowest in the < order) building all the clusters of client nodes that include 1, satisfying condition 1). Then, if all possible clusters up to i = k have been obtained, consider i = k +1 (where, of course, k < k + 1). All routes starting and ending in a node of N^s that go through client nodes in $N^c - \{1, \cdots, k\}$, including node k + 1, satisfying condition 1), are constructed.

For each $i \in N^c$ its class of clusters, denoted S^i, is endowed with a weak order Đi (reflexive, antisymmetric and transitive, i.e. allowing ties) according to the number of clients in each cluster. Among clusters covering the same number of client nodes, say $\{1, \cdots, i, \cdots, k\}$ and $\{1, \cdots, i, \cdots, l\}$, if k

< l then $\{1,\cdots,i,\cdots,k\}$ Đi $\{1,\cdots,i,\cdots,l\}$. The highest rank in Đi corresponds to routes covering i as well as the largest possible number of other client nodes in the upper level of order <. As an example, consider the case of 10 client nodes demanding 50 units and a vehicle capacity of 100 units. **Table 1** shows the distances among nodes, being 0 the storage site from which departs the vehicle. **Table 2** shows the result of the first stage of the procedure. Notice that, since the capacity of the vehicle is 100 and each client node demands 50, a route that clusters some clients together can only consist of two of them. So, for instance the group corresponding to node 9 includes only two possible clusters, {9} and {9,10}, because all other nodes have already been assigned to other clusters. Cluster {9,10} is of a higher order than {9} because it includes two clients instead of just one node.

3.2. Second Stage

At this point a Genetic Algorithm [24] is applied, designed for its use in this problem. An individual is codified in a chromosome constructed in such way as to satisfy conditions 2) and 3) and therefore are feasible under problem (2). The coding operates on a lexicographic basis. Each $i \in N^c$ is given a number, starting with i = 1, and following the process according to order <. So, take all the clusters ordered according to Đ1. This order can be represented by the linear numerical order $1,\cdots,|S^1|$. So, if the k-th cluster in the ordering is chosen, the corresponding gene in the chromosome is k. For i = j, the order Đj is applied to a subset of S^j obtained by eliminating ("filtered out") all the clusters that include nodes already present in clusters corresponding to genes i = 1,···, j – 1. The new ordering is denoted $\angle i'$. Then the j-th gene in the chromosome is the rank in $\angle i'$ of the cluster that is chosen. If for a given i the domain obtained is Æ (because all possible nodes belong to clusters corresponding to genes 1 to i – 1), any number between 1 and $|S^i|$ can be assigned to the i-th gene. In this way, a chromosome is defined, each corresponding to a feasible solution to problem (2). The Genetic Algorithm seeks to find the optimal one.

Table 1. Distances—Case: 10 nodes / Demand per node: 50 units / Vehicle capacity: 100 units.

Nodes	0	1	2	3	4	5	6	7	8	9	10
0	0.00	5.00	14.14	7.07	11.18	15.81	14.14	5.00	11.18	11.18	18.03
1	5.00	0.00	11.18	11.18	15.81	20.62	18.03	10.00	7.07	10.00	22.36
2	14.14	11.18	0.00	15.81	25.00	25.50	20.00	18.03	5.00	20.62	32.02
3	7.07	11.18	15.81	0.00	11.18	10.00	7.07	5.00	15.00	18.03	18.03
4	11.18	15.81	25.00	11.18	0.00	11.18	15.00	7.07	22.36	15.81	7.07
5	15.81	20.62	25.50	10.00	11.18	0.00	7.07	11.18	25.00	25.00	15.00
6	14.14	18.03	20.00	7.07	15.00	7.07	0.00	11.18	20.62	25.00	20.62
7	5.00	10.00	18.03	5.00	7.07	11.18	11.18	0.00	15.81	14.14	14.14
8	11.18	7.07	5.00	15.00	22.36	25.00	20.62	15.81	0.00	15.81	29.15
9	11.18	10.00	20.62	18.03	15.81	25.00	25.00	14.14	15.81	0.00	20.00
10	18.03	22.36	32.02	18.03	7.07	15.00	20.62	14.14	29.15	20.00	0.00

Table 2. Clusters—Case: 10 nodes / Demand per node: 50 units / Vehicle capacity: 100 units.

Nodes	Clusters									
0	.	.	.	.	.	.	.	.	.	.
1	1	1.2	1.3	1.4	1.5	1.6	1.7	1.8	1.9	1.10
2	2	2.3	2.4	2.5	2.6	2.7	2.8	2.9	2.10	.
3	3	3.4	3.5	3.6	3.7	3.8	3.9	3.10	.	.
4	4	4.5	4.6	4.7	4.8	4.9	4.10	.	.	.
5	5	5.6	5.7	5.8	5.9	5.10	.	.	.	.
6	6	6.7	6.8	6.9	6.10	.	.	.	.	.
7	7	7.8	7.9	7.10	.	.	.	.	.	.
8	8	8.9	8.10	.	.	.	.	.	.	.
9	9	9.10	.	.	.	.	.	.	.	.
10	10	.	.	.	.	.	.	.	.	.

We can see some examples of how chromosomes encode clusters in **Table 3**, (3-3-2-1-4-2-3-2-1-1) and **Table 4** (5-2-1-5-1-4-2-2-1-1): all routes start and end at node 0 and go through two client nodes; in gray we see the filtered out clusters and in red the selection. Consider for instance the second gene in **Table 3**: it has to discard (written in gray) all routes that go through node 3, since it has already been assigned to the first gene. Number 3 assigned to this gene indicate that the route selected is the third in the list of non-filtered routes. On the other hand, a number assigned to a gene with a row with all routes filtered out does not have any effect. The initial population consists of individuals whose chromosomes are constituted by genes selected at random. Once established a chromosome, it is decoded and evaluated. This evaluation is performed according to problem (2). For the computation of each cluster we use Lin-Kerninghan's algorithm that yields the minimal distance among its nodes. The following operators have been chosen in the light of some previous evaluations. The cross-over operator is the two-point one (**Table 5**, Parent 1: 3-3- 2-_-_-_-_-2-1-1, Parent 2: _-_-_-5-1-4-2-_-_-_, Offspring: 3-3-2-5-1-4-2-2-1-1) and the mutation operator alters gene values at 10% of the chromosome (Mutation (**Table 6**, Offspring*: 3-3-2-5-1-2-2-2-1-1)). Finally, the selection is by ranking.

The Genetic Algorithm runs until the solution gets stabilized. These tests were implemented with a C++ program running on a 3.00 GHZ CPU with a 2.00 GB RAM. The parameters of the experiments were as follows. The size of the population: 100; the cross-over probability: 0.70 and the probability of mutation: 0.01.

Before running the algorithm over problems in the literature we experimented over this example. **Figure 1** shows the nodes and their proportional distances in a two-dimensional framework. Figures 2 to 7 show the routes generated in several generations and the minimal total distance covered by the selected routes.

Table 3. Example of a chromosome (3-3-2-1-4-2-3-2-1-1) that will act as parent 1 in the cross-over of **Table 5**.

Chr.	Nodes	Clusters									
3	1	0-1-0	0-1-2-0	0-1-3-0	0-1-4-0	0-1-5-0	0-1-6-0	0-1-7-0	0-1-8-0	0-1-9-0	0-1-10-0
3	2	0-2-0	0-2-3-0	0-2-4-0	0-2-5-0	0-2-6-0	0-2-7-0	0-2-8-0	0-2-9-0	0-2-10-0	-
2	3	0-3-0	0-3-4-0	0-3-5-0	0-3-6-0	0-3-7-0	0-3-8-0	0-3-9-0	0-3-10-0	-	-
1	4	0-4-0	0-4-5-0	0-4-6-0	0-4-7-0	0-4-8-0	0-4-9-0	0-4-10-0	-	-	-
4	5	0-5-0	0-5-6-0	0-5-7-0	0-5-8-0	0-5-9-0	0-5-10-0	-	-	-	-
2	6	0-6-0	0-6-7-0	0-6-8-0	0-6-9-0	0-6-10-0	-	-	-	-	-
3	7	0-7-0	0-7-8-0	0-7-9-0	0-7-10-0	-	-	-	-	-	-
2	8	0-8-0	0-8-9-0	0-8-10-0	-	-	-	-	-	-	-
1	9	0-9-0	0-9-10-0	-	-	-	-	-	-	-	-
1	10	0-10-0	-	-	-	-	-	-	-	-	-

Table 4. Example of a chromosome (5-2-1-5-1-4-2-2-1-1) that will act as parent 2 in the cross-over of **Table 5**.

Chr.	Nodes	Clusters									
5	1	0-1-0	0-1-2-0	0-1-3-0	0-1-4-0	0-1-5-0	0-1-6-0	0-1-7-0	0-1-8-0	0-1-9-0	0-1-10-0
2	2	0-2-0	0-2-3-0	0-2-4-0	0-2-5-0	0-2-6-0	0-2-7-0	0-2-8-0	0-2-9-0	0-2-10-0	-
1	3	0-3-0	0-3-4-0	0-3-5-0	0-3-6-0	0-3-7-0	0-3-8-0	0-3-9-0	0-3-10-0	-	-
5	4	0-4-0	0-4-5-0	0-4-6-0	0-4-7-0	0-4-8-0	0-4-9-0	0-4-10-0	-	-	-
1	5	0-5-0	0-5-6-0	0-5-7-0	0-5-8-0	0-5-9-0	0-5-10-0	-	-	-	-
4	6	0-6-0	0-6-7-0	0-6-8-0	0-6-9-0	0-6-10-0	-	-	-	-	-
2	7	0-7-0	0-7-8-0	0-7-9-0	0-7-10-0	-	-	-	-	-	-
2	8	0-8-0	0-8-9-0	0-8-10-0	-	-	-	-	-	-	-
1	9	0-9-0	0-9-10-0	-	-	-	-	-	-	-	-
1	10	0-10-0	-	-	-	-	-	-	-	-	-

Table 5. Cross-over (offspring: 3-3-2-5-1-4-2-2-1-1).

Chr.	Nodes	Clusters									
3	1	0-1-0	0-1-2-0	0-1-3-0	0-1-4-0	0-1-5-0	0-1-6-0	0-1-7-0	0-1-8-0	0-1-9-0	0-1-10-0
3	2	0-2-0	0-2-3-0	0-2-4-0	0-2-5-0	0-2-6-0	0-2-7-0	0-2-8-0	0-2-9-0	0-2-10-0	-
2	3	0-3-0	0-3-4-0	0-3-5-0	0-3-6-0	0-3-7-0	0-3-8-0	0-3-9-0	0-3-10-0	-	-
5	4	0-4-0	0-4-5-0	0-4-6-0	0-4-7-0	0-4-8-0	0-4-9-0	0-4-10-0	-	-	-
1	5	0-5-0	0-5-6-0	0-5-7-0	0-5-8-0	0-5-9-0	0-5-10-0	-	-	-	-
4	6	0-6-0	0-6-7-0	0-6-8-0	0-6-9-0	0-6-10-0	-	-	-	-	-
2	7	0-7-0	0-7-8-0	0-7-9-0	0-7-10-0	-	-	-	-	-	-
2	8	0-8-0	0-8-9-0	0-8-10-0	-	-	-	-	-	-	-
1	9	0-9-0	0-9-10-0	-	-	-	-	-	-	-	-
1	10	0-10-0	-	-	-	-	-	-	-	-	-

Table 6. Mutation (offspring*: 3-3-2-5-1-4-2-2-1-1).

Chr.	Nodes	Clusters									
3	1	0-1-0	0-1-2-0	0-1-3-0	0-1-4-0	0-1-5-0	0-1-6-0	0-1-7-0	0-1-8-0	0-1-9-0	0-1-10-0
3	2	0-2-0	0-2-3-0	0-2-4-0	0-2-5-0	0-2-6-0	0-2-7-0	0-2-8-0	0-2-9-0	0-2-10-0	-
2	3	0-3-0	0-3-4-0	0-3-5-0	0-3-6-0	0-3-7-0	0-3-8-0	0-3-9-0	0-3-10-0	-	-
5	4	0-4-0	0-4-5-0	0-4-6-0	0-4-7-0	0-4-8-0	0-4-9-0	0-4-10-0	-	-	-
1	5	0-5-0	0-5-6-0	0-5-7-0	0-5-8-0	0-5-9-0	0-5-10-0	-	-	-	-
2	6	0-6-0	0-6-7-0	0-6-8-0	0-6-9-0	0-6-10-0	-	-	-	-	-
2	7	0-7-0	0-7-8-0	0-7-9-0	0-7-10-0	-	-	-	-	-	-
2	8	0-8-0	0-8-9-0	0-8-10-0	-	-	-	-	-	-	-
1	9	0-9-0	0-9-10-0	-	-	-	-	-	-	-	-
1	10	0-10-0	-	-	-	-	-	-	-	-	-

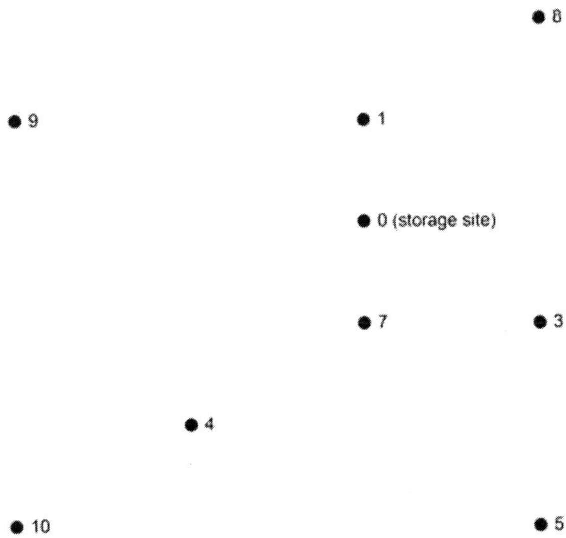

Figure 1. Topology of the problem.

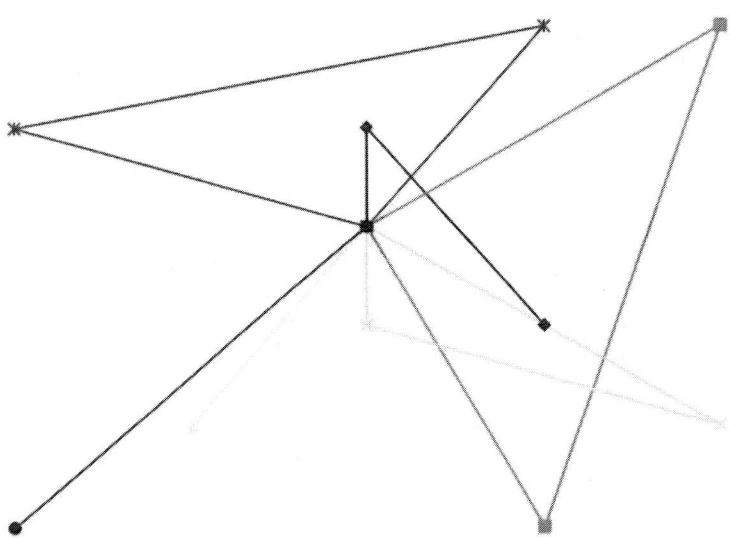

Figure 2. Generation 10, distance: 205.61.

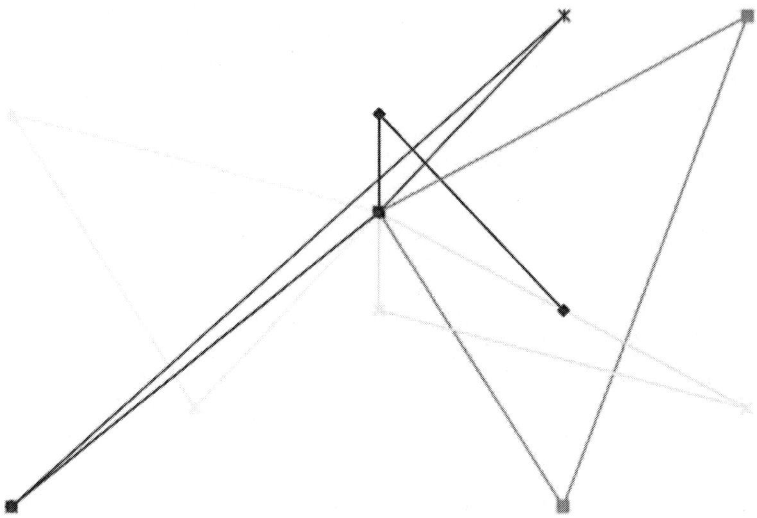

Figure 3. Generation 20, distance: 205.56.

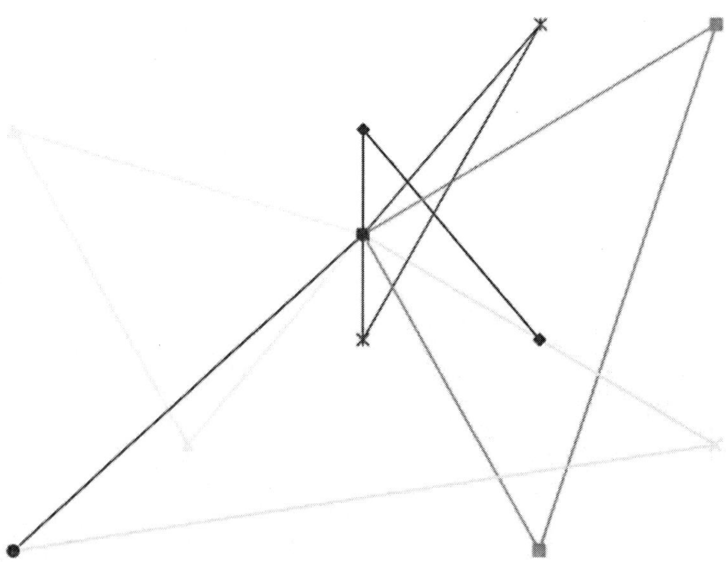

Figure 4. Generation 45, distance: 201.65.

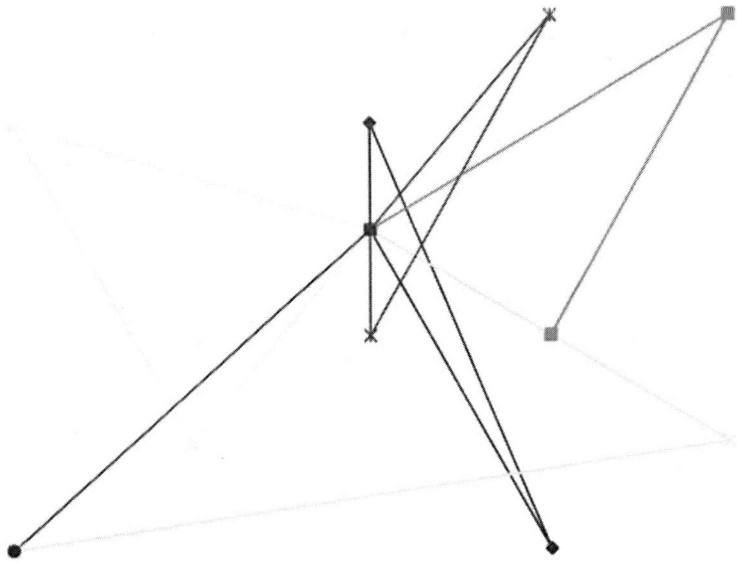

Figure 5. Generation 75, distance: 201.40.

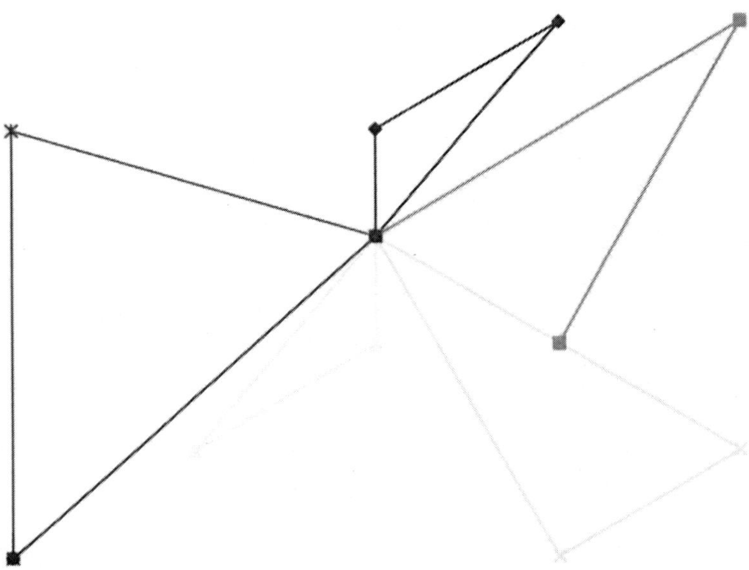

Figure 6. Generation 100, distance: 169.76.

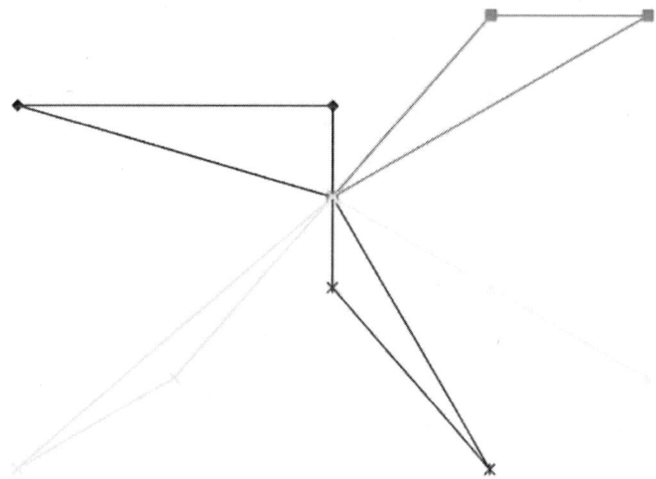

Figure 7. Generation 150, distance: 153.06.

4. EXPERIMENTS AND RESULTS

Van Breedam [25] presents a large number of instances of features and values. He suggested also a reduced set of representative cases to be solved by different heuristics. Some of them involve 100 nodes with a total demand of 1000 units. We follow here this convention and use the author's nomenclature for the cases with the same capacity of vehicles. The problems tested range from Bre-1 to Bre-15. From Bre-1 to Bre-6 only the capacity of vehicles is restricted. The demand on each client node is 10 units, while the vehicle capacity is 100 units in Bre-1 and Bre-2, 50 units in Bre-3 and Bre-4, while it is of 200 units in Bre-5 and Bre-6. Problems Bre-7 and Bre-8 have not been considered because they involve both deliveries and pickups. In most of these problems a single type of good is assumed. Exceptions are problems Bre-9 to Bre-11, in which demands are heterogeneous. Finally Bre-12 to Bre-15 are problems involving time windows, and therefore are not included in our experiments. **Table 7** presents the results of our tests. The best known solution, drawn from [25,26], is presented as well as information on the solutions under our procedure (best solution, average evaluation, success, mean solution, average running times and percentage of difference with the best known solution). With respect to the processing time, some instances took only minutes while others required hours, although in average the best solutions took not much longer than an hour to be found.

Table 7. Results for the problems of Van Breedam [25,26].

Instance	Best Known Solution	Best Solution Found					
		Best Solution	Evaluations	Success (%)	Average Solution	Δ (%)	Average Time (min)
Bre-1	1106.00	1157.00	395347.12 ± 125177.89	42.00	1169.02 ± 14.76	−4.61	25.16
Bre-2	1506.00	1587.00	374623.02 ± 25544.61	18.00	1595.75 ± 16.91	−5.38	24.11
Bre-3	1751.00	1751.00	121640.67 ± 25684.27	100.00	1751.00 ± 0.00	0.00	11.20
Bre-4	1470.00	1470.00	287936.32 ±14739.13	100.00	1470.00 ± 0.00	0.00	19.68
Bre-5	950.00	985.00	326733.84 ± 112846.01	76.00	994.12 ± 11.49	−3.68	21.66
Bre-6	969.00	979.00	523967.47 ± 162502.07	62.00	982.43 ± 6.59	−1.03	31.72
Bre-9	1690.00	1790.00	657281.22 ± 21838.39	17.00	1797.93 ± 8.03	−5.92	38.52
Bre-10	1026.00	1026.00	871531.96 ± 326830.70	52.00	1035.85 ± 9.42	0.00	49.45
Bre-11	1128.00	1187.00	1342635.29 ± 543765.65	100.00	1187.00 ± 0.00	−5.23	73.47

5. CONCLUSION

We presented a new algorithm for the CVRP (Capacitated Vehicle Routing Problem), distinguishing between the standard and its alternative formulation. The details of the procedure were described and the comparison with results in the literature has been presented. While the outcomes are similar to them, their running times are not. In experiments not discussed here we noticed that problems that exceed 100 nodes the processing time is much longer, due to the demands of the first stage of the algorithm. We plan to add further search filters at the optimization stage as well as to try with other meta-heuristics to run on the alternative model.

6. ACKNOWLEDGEMENTS

This work has been supported by several funding sources. The authors are in particular thankful to the National Research Council of Argentina (CONICET) as well as to the Universidad Nacional del Sur project PGI 24/J056.

REFERENCES

1. D. Ambrosino and A. Sciomachen, "A Food Distribution Network Problem: A Case Study," IMA Journal of Management Mathematics, Vol. 18, No. 1, 2007, pp. 33-53.
2. L. D. Bodin and B. L. Golden, "Classification in Vehicle Routing and Scheduling," Networks, Vol. 11. No. 2, 1981, pp. 97-108.
3. S. N. Kumar and R. Panneerselvam, "A Survey on the Vehicle Routing Problem and Its Variants," Intelligent Information Management, Vol. 4, No. 3, 2012, pp. 66-74.
4. G. Laporte and I. H. Osman, "Routing Problems: A Bibliography," Annals of Operations Research, Vol. 61, No. 1, 1995, pp. 227-262.
5. E. Mota, V. Campos and A. Corbéran, "A New Metaheuristic for the Vehicle Routing Problem with Split Demands," In: J. van Hemert and C. Cotta, Eds., Evolutionary Computation in Combinatorial Optimization, Lecture Notes in Computer Science Vol. 4446, Springer-Verlag, Berlin, 2007, pp. 121-129.
6. R. Tavakkoli-Moghaddam, N. Safaei, M. Kah and M. Rabbani, "A New Capacitated Vehicle Routing Problem with Split Service for

Minimizing Fleet Cost by Simulated Annealing," Journal of the Franklin Institute, Vol. 344, No. 5, 2007, pp. 406-425.

7. J. M. Belenguer, E. Benavent, N. Labadi, C. Prins and M. Reghioui, "Split Delivery Capacitated Arc-Routing Problem: Lower Bound and Metaheuristic," Transportation Science, Vol. 44, No. 2, 2010, pp. 206-220.

8. S. R. Thangiah, A. Fergany and S. Awan, "Real-Time Split-Delivery Pickup and Delivery Time Window Problems with Transfers," Central European Journal of Operations Research, Vol. 15, No. 4, 2007, pp. 329-349.

9. M. Desrochers, J. Desrosiers and M. M. Solomon, "A New Optimization Algorithm for the Vehicle Routing Problem with Time Windows," Operations Research, Vol. 40, No. 2, 1002, pp. 342-354.

10. C. G. Lee, M. A. Epelman, C. White and Y. A. Bozer, "A Shortest Path Approach to the Multiple-Vehicle Routing Problem with Split Pick-Ups," Transportation Research, Vol. 40, No. 4, 2006, pp. 265-284.

11. P. Belfiore and H. T. I. Yoshizaki, "Scatter Search for a Real-Life Heterogeneous Fleet Vehicle Routing Problem with Time Windows and Split Deliveries in Brazil," European Journal of Operational Research, Vol. 199, No. 3, 2009, pp. 750-758.

12. J. H. Wilck IV and T. M. Cavalier, "A Construction Heuristic for the Split Delivery Vehicle Routing," American Journal of Operations Research, Vol. 2, No. 2, 2012, pp. 153-162.

13. Y. Nagao and H. Nagamochi, "A DP-Based Heuristic Algorithm for the Discrete Split Delivery Vehicle Routing Problem," Journal of Advanced Mechanical Design, Systems, and Manufacturing, Vol. 1, No. 2, 2007, pp. 217-226.

14. Y. Yu, H. Chen and F. Chu, "A New Model and Hybrid Approach for Large Scale Inventory Routing Problems," European Journal of Operational Research, Vol. 189, No. 3, 2008, pp. 1022-1040.

15. M. A. Nowak, O. Ergun and C. C. White, "An Empirical Study on the Benefit of Split Loads with the Pickup and Delivery Problem," European Journal of Operational Research, Vol. 198, No. 3, 2009, pp. 734-740.

16. Z. Shen, M. M. Dessouky and F. Ordóñez, "A Two-Stage Vehicle Routing Model for Large-Scale Bioterrorism Emergencies," Networks, Vol. 54, No. 4, 2009, pp. 255-269.

17. L. Moreno, M. P. de Aragao and E. Uchoa, "Improved Lower Bounds for the Split Delivery Vehicle Routing Problem," Operations Research Letters, Vol. 38, No. 4, 2010, pp. 302-306.

18. M.-C. Bolduc, G. Laporte, J. Renaud and F. F. Boctor, "A Tabu Search Heuristic for the Split Delivery Vehicle Routing Problem with Production and Demand Calendars," European Journal of Operational Research, Vol. 202, No. 1, 2010, pp. 122-130.

19. U. Derigs, B. Li and U. Vogel, "Local Search-Based Metaheuristics for the Split Delivery Vehicle Routing Problem," Journal of the Operational Research Society, Vol. 61, No. 9, 2009, pp. 1356-1364.

20. M. Jin, K. Liu and R. O. Bowden, "A Two-Stage Algorithm with Valid Inequalities for the Split Delivery Vehicle Routing Problem," International Journal of Production Economics, Vol. 105, No. 1, 2007, pp. 228-242.

21. R. E. Aleman and R. R. Hill, "A Tabu Search with Vocabulary Building Approach for the Vehicle Routing Problem with Split Demands," International Journal of Metaheuristics, Vol. 1, No. 1, 2010, pp. 55-80.

22. R. E. Aleman, X. Zhang and R. R. Hill, "An Adaptive Memory Algorithm for the Split Delivery Vehicle Routing Problem," Journal of Heuristics, Vol. 16, No. 3, 2010, pp. 441-473.

23. S. Basu, "Tabu Search Implementation on Traveling Salesman Problem and Its Variations: A Literature Survey," American Journal of Operations Research, Vol. 2, No. 2, 2012, pp. 163-173.

24. D. Goldberg, "Genetic Algorithms in Search, Optimization, and Machine Learning," Addison-Wesley Longman Publishing Co., Inc., Boston, 1989.

25. A. Van Breedam, "An Analysis of the Behavior of Heuristics for the Vehicle Routing Problem for a Selection of Problems with Vehicle Related, Customer-Related, and Time-Related Constraints," Ph.D. Thesis, University of Antwerp—RUCA, Antwerpen, 1994.

26. E. Alba and B. Dorronsoro, "A Hybrid Cellular Genetic Algorithm for the Capacitated Vehicle Routing Problem," In: Engineering Evolutionary Intelligent Systems, Chapter 13, Studies in Computational Intelligence, SpringerVerlag, Heidelberg, 2008, pp. 379-422.

2

CHAPTER

DESIGN OF TYPE-1 AND INTERVAL TYPE-2 FUZZY PID CONTROL FOR ANESTHESIA USING GENETIC ALGORITHMS

Hugo Araujo, Bo Xiao, Chuang Liu, Yanbin Zhao, H. K. Lam

Department of Informatics, King's College London, London, UK

ABSTRACT

This paper presents the automatic drug administration for the regulation of bispectral (BIS) index in the anesthesia process during the clinical surgery by controlling the concentration target of two drugs, namely, propofol and remifentanil. To realize the automatic drug administration, real clini- cal data are collected for 42 patients for the construction of patients' models consisting of pharma- cokinetic and pharmacodynamic models describing the dynamics reacting to the input drugs. A nominal anesthesia model is obtained by taking the average of 42 patients' models for the design of control scheme. Three PID controllers are employed, namely linear PID controller, type-1 (T1) fuzzy PID controller and interval type-2 (IT2) fuzzy PID controller, to regulate the BIS index using the nominal patient's model. The PID gains and membership functions are obtained using genetic algorithm (GA) by minimizing a cost function measuring the control performance. The best train- ed PID controllers are tested under different scenarios and compared in terms of control performance. Simulation results show that the IT2 fuzzy PID controller offers the best control strategy regulating the BIS index while the T1 fuzzy PID controller comes the second.

Keywords: Anesthesia, Propofol, Remifentanil, BIS, PID, Fuzzy, Control, Simulation, Close-Loop

1. INTRODUCTION

Anesthesia is a reversible state of people who are temporarily lack of consciousness in the purpose of undergo- ing a surgery without pain. General process of anesthesia consists of induction, maintenance, emergence, and re- covery [1] . The depth of anesthesia (DOA) represents the level of consciousness [2] , which is to be controlled through the process of anesthesia for varieties of control targets such as steady error, settling time, and overshoot. In general, control strategies can be categorized into two classes: open-loop control and closed-loop control. In open-loop control, based on the knowledge and experience, anesthetists manually adjust the drug dosage to maintain the DOA assisted by some clinical indices of patients. In closed-loop control, the drug dosage is auto- matically adjusted according to some indices of DOA, which makes control input continuous and responsive [3] . Closed-loop control is also expected to avoid over- and under-dosage and suppress the adverse effect of inter-individual differences [4] . Despite of its advantages, the stability of closed-loop control needs to be ensured due to the automated process without supervision [5] .

To achieve the stabilization based on closed-loop control, the following components are required [6] : 1) a patient model, with an output as the index of DOA; 2) a controller for stabilization, such as proportional-integral-differential (PID) controllers. Therefore, to begin with, an estimated patient model is required to represent real patients supporting the control design and performance evaluation. A widely employed mathematical model consists of a linear pharmacokinetic (PK) model and a nonlinear pharmacodynamic (PD) model [7] . The PK mo- del uses the drug dosage as the input and the drug concentration as the output. In what follows, the PD model exploits the drug concentration as the input and exports the index of DOA.

As stated above, the control input of the overall patient model is the drug dosage. According to the types of drugs, the anesthesia can be classified to inhalational and intravenous anesthesia. Inhalational anesthesia has been used since the mid-19th century with for instance nitrous oxide, isoflurane, and halothane as commonly used drugs [8] . In contrast, intravenous anesthesia with propofol and remifentanil as commonly used drugs [9] has the following advantages [10] : separate provision of anesthesia from ventilation, reduced atmospheric pollution, rapid and clear-headed recovery, and so on. While inhalational anesthesia is still frequently used for children, in- travenous anesthesia becomes more and more popular due to the rapid and safe transition [11] .

Main difficulties involved in modeling are the determination of PD model parameters and the selection of the output index. In general, the parameters of PK and PD models need to be determined beforehand. For the PK mo- del, parameters can be estimated depending on sex, age, and weight of patients. Nonetheless, for the PD model, it is not possible to estimate the parameters for certain patient. Accordingly, it requires the controller to be robust in a domain of PD model parameters [12] . Although PK model parameters are varied with patients and PD model parameters are even changing for one patient, a general index can be designed to evaluate the DOA for all patients [13] . The bispectral (BIS) index [14] is an extensively accepted index to measure the DOA. It is a uni- variate dimensionless parameter from 0 to 100, which is devised from a large set of electroencephalogram (EEG) data [15] .

With regard to the controllers for stabilization, several control problems are faced for the closed-loop control of anesthesia: stability, which is the basic control objective; robustness, which overcomes the uncertainty of PD model parameters, measurement noise, surgical stimulation, and so forth; adaptiveness, which makes the controller adaptive to different patients rather than only one patient. Commonly used controllers include PID controllers, model-based controllers, and knowledge-based controllers [6] .

To close the loop of control systems, classical PID controller is a doubtless option due to its successful applications in other areas. For stabilization, it is necessary to tune the parameters of PID controllers for a certain patient model. While the tuning process is an empirical manner, some tuning schemes were even developed to gua- rantee the robustness [16] [17] . Combined with patient model identification from the induction phase of anesthe- sia [18] , the adaptive ness can be further achieved. In addition, combined with genetic algorithm (GA), the para- meters of PID controllers can also be online optimized [19] .

Despite of various causes of the changes in PK and PD model parameters, the model-based controller relies on the current model which reflects the patient's current pharmacological behavior. For this reason, the patient model needs to be updated through the overall process of anesthesia. This online adaptation can be achieved by several approaches such as Kalman filter algorithm [20] , Bayesian-based adaptive control [21] , and adaptive ge- netic fuzzy clustering algorithm [19] . Alternatively, the variability of PK and PD models between individuals as well as surgical stimulation and anesthetic-analgesic interaction can be explicitly considered offline such that the stability is mathematically guaranteed [22] . Additionally, a Lyapunov-based adaptive controller was developed to attain partial asymptotic regulation [23] .

Unlike model-based controllers, knowledge-based controllers do not require a known mathematical model. In fuzzy logic controllers, for example, decisions are made based on fuzzy rules predefined by expert knowledge and experience. Compared with linear PID controllers and model-based controller, knowledge-based controllers are easier to implement without tuning PID parameters or mathematical derivation. A hybrid control scheme, namely fuzzy-PID control, was developed to combine the merits of both control strategies [24] . Moreover, a multivariable neural-fuzzy controller was proposed to simultaneously administrate both propofol and remifentanil [25] . However, one problem of knowledge-based controllers is that the interaction of each piece of knowledge makes the controller less transparent and makes it difficult to achieve adaptive control [6] . To deal with this problem, a direct adaptive interval type-2 (IT2) fuzzy logic controller was proposed for multivariable anesthesia systems [26] and a genetic fuzzy logic controller was developed to adjust fuzzy rules using GA [19].

Due to the advantages of closed-loop control, in this paper, we aim to implement the closed-loop control for anesthesia model. The performance of controllers is ensured by simulation and optimization. For the anesthesia model used in this paper, while existing anesthesia model is utilized, the parameters of PD model are specially obtained from clinical data collected from 42 patients. Following the same design procedure, these tailor-made controllers can be redesigned for other patients as long as clinical data have been collected. In the anesthesia model, the co-administration of both propofol and remifentanil is investigated. To realize the drug administration, we propose two groups of strategies: two fuzzy PID controllers and one fuzzy PID controller with scaling factors. For each group of strategies, linear PID controller, type-1 (T1) fuzzy PID controllers and IT2 fuzzy PID controllers are designed. To draw a distinction from existing IT2 fuzzy PID controllers, we combine IT2 fuzzy PID controllers with GA. All PID gains, scaling factors and parameters of membership functions are optimized by GA in an offline manner subject to a performance index (cost function) which quantifies the performance of the controllers. A BIS training profile which sets different local targets for regulation considering the real anesthesia situation is employed for training purposes. The trained PID control strategies will be tested by a testing profile to verify their performance towards unseen working conditions. Comparisons are made among all controllers to demonstrate the characteristics of each control strategy.

Following sections are organized in this sequence: in Section 2, a multivariable anesthesia model used in this paper is described; in Section 3, the control background of linear PID controller, T1 fuzzy PID controllers and IT2 fuzzy PID controllers is introduced; in Section 4, we design all the control strategies and overall procedure and necessary information of the simulation; in Section 5, simulation results are provided and comparison and analysis are carried out; finally in Section 6, a conclusion is drawn.

2. MULTIVARIABLE ANESTHESIA MODELING

In this section, the anesthesia model with two drugs is introduced. As shown in Figure 1, the anesthesia model consists of the target controlled infusion (TCI) system, the PK model, and the PD model, where Cpt_{prop} and Cpt_{remi} are plasmatic concentration targets, I_p and I_r are infusion rates, Cp_{prop} and Cp_{remi} are plasmatic concentrations, Ce_{prop} and Ce_{remi} are effect-site concentrations, and the subscript $prop$ and $remi$ are for propofol and remifentanil, respectively. Details of each module are described in the following subsections. The clinical protocol for obtaining the parameters of PD model is also described.

2.1. Pharmacokinetic Modeling

PK model is applied to reckon the relation from the drug infusion rate and the amount of drugs to plasmatic concentration (Cp) and effect-site concentration (Ce). In this paper, we utilize the PK model from Marsh [27] and Minto [28] for propofol and remifentanil, respectively. A three-compartmental structure of these PK models is presented in Figure 2 which can be formulated by the following dynamic system:

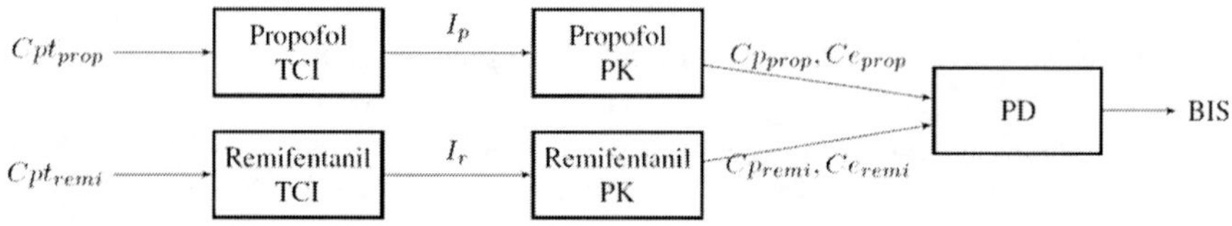

Figure 1. A block diagram of multivariable anesthesia model.

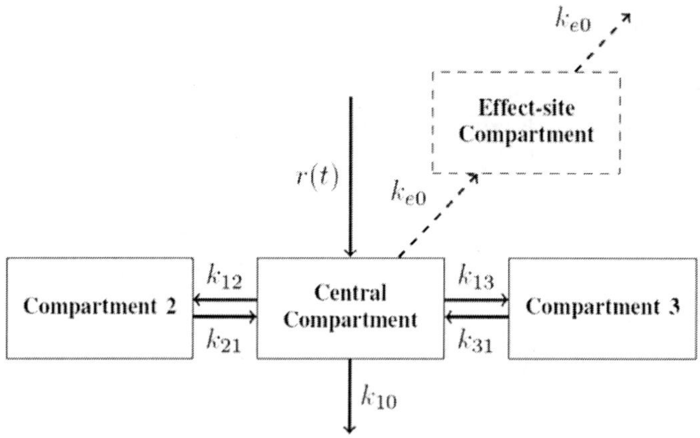

Figure 2. Structure of a three-compartmental PK model with effect-site com- partment [29].

$$\dot{x}(t) = Fx(t) + Gr(t), \quad y(t) = Cx(t), \tag{1}$$

where $y(t) = \begin{bmatrix} C_p(t) \\ C_e(t) \end{bmatrix}$, $x(t) = \begin{bmatrix} m_1(t) \\ m_2(t) \\ m_3(t) \\ C_e(t) \end{bmatrix}$, $F = \frac{1}{60}\begin{bmatrix} -k_{10} - k_{12} - k_{13} & k_{21} & k_{31} & 0 \\ k_{12} & -k_{21} & 0 & 0 \\ k_{13} & -k_{31} & 0 & 0 \\ \frac{k_{e0}}{V_1} & 0 & 0 & -k_{e0} \end{bmatrix}$,

$G = \frac{1}{60}\begin{bmatrix} 1 \\ 0 \\ 0 \\ 0 \end{bmatrix}$, $C = \begin{bmatrix} \frac{1}{V_1} & 0 & 0 & 0 \\ 0 & 0 & 0 & 1 \end{bmatrix}$;

$t(s)$ is the continuous time; $r(t)(\mathrm{mg/min})$ is the drug infusion rate; $k_{ij}(\mathrm{min}^{-1})$ is the rate constant for dis- tribution and elimination of the anesthetic drug; $V_1(t)$ is the volume of central compartment; and $k_{e0}(\mathrm{min}^{-1})$ is a scalar describing the delay on time of the plasmatic-effect-site drug concentration equilibration.

2.2. Pharmacodynamic Modeling

After Cp and Ce are obtained from PK model, PD model maps Ce to the BIS value which reflects the anesthetic condition of patients. In general, PD model is established by a sigmoid $E_{\max}$ model named as Hill curve for single drug. Based on the Hill curve, a response surface model was proposed by Minto et al. [30] to characterize drugs' interaction for multi-drugs. The PD model for representing propofol and remifentanil can be formulated as follows [31]:

$$\text{Effect}\ \left(Ce_{prop}, Ce_{remi}\right) = E_0 + \left(E_{\max} - E_0\right)\frac{\left(\dfrac{U_{prop}+U_{remi}}{U_{50}(\theta)}\right)^{\gamma}}{1+\left(\dfrac{U_{prop}+U_{remi}}{U_{50}(\theta)}\right)^{\gamma}}, \tag{2}$$

where E_0 is the clinical effect when no drug is infused; $E_{\max}$ is the maximum drug effect; γ is the steepness of the concentration-response curve; θ is the ratio of the two drugs defined in Equation (3); U_{prop} and U_{remi} are the normalized Ce_{prop} and Ce_{remi}, respectively, defined in Equation (4); Ce_{50} is the effect-site drug concentration associated with 50% maximal drug potency; $U_{50}(\theta)$ is defined in Equation (5); β is a predefined scalar.

$$\theta = \frac{U_{prop}}{U_{prop} + U_{remi}} \tag{3}$$

$$U_{prop} = \frac{Ce_{prop}}{Ce_{50,prop}} \qquad U_{remi} = \frac{Ce_{remi}}{Ce_{50,remi}} \tag{4}$$

$$U_{50}(\theta) = 1 - \beta\theta + \beta\theta^2 \tag{5}$$

In this paper, let us define E_0 and E_{max} as 98 and 0, respectively. $Ce_{50,prop}$, $Ce_{50,remi}$, β and γ are patient-dependent parameters which are estimated by clinical data explained in the following subsection.

2.3. Target Controlled Infusion System

TCI system transfers the plasmatic concentration targets (Cpt) to corresponding drug infusion rates. Practically, the infusion can be realized by Alaris PK syringe pumps which incorporates the TCI system. These pumps are able to infuse adequate amount of drugs according to predefined Cpt. Furthermore, they can predict the realistic concentration based on PK model.

The TCI system consists of a bolus-elimination-transfer (BET) infusion scheme designed from the PK model so the Cpt is achieved. The initial bolus, loading dose (LD), is calculated according to Equation (6) where $V_1(t)$ is the volume of central compartment represented on Figure 2. The LD bolus will ensure the Cpt is achieved however it needs to be followed by a continuous infusion, $r(t)$, defined by Equation (7) to ensure Cpt is maintained due to elimination and compartmental transfer of the drug [32] .

$$LD = V_1 Cpt, \tag{6}$$

$$r(t) = LD\left(k_{10} + k_{12}e^{-k_{21}t} + k_{13}e^{-k_{31}t}\right), \tag{7}$$

where $t(s)$ is the continuous time; $k_{ij}\left(\min^{-1}\right)$ is the rate constant for distribution and elimination of the anesthetic drug defined in Section 2.1.

TCI systems are also able to deal with Cpt changes. When Cpt increases an initial additional loading dose (ADDLD), Equation (6) and Equation (7) will become the following equations to maintain the new $Cpt(NCpt)$.

$$ADDLD = V_1\left(NCpt - Cpt\right), \tag{8}$$

$$r(t) = V_1 NCpt\left(k_{10} + k_{12}e^{-k_{21}(t-\tau)} + k_{13}e^{-k_{31}(t-\tau)}\right) - \left(k_{21}m_2(\tau)e^{-k_{21}(t-\tau)} + k_{31}m_3(\tau)e^{-k_{31}(t-\tau)}\right), \tag{9}$$

where τ is the time of Cpt change; $k_{ij}\left(\min^{-1}\right)$ are the rate constant for distribution and elimination of the anesthetic drug; $m_2(\tau)(\text{mg})$ is the amount of anesthetic drug in the second compartment at the time of target change; $m_3(\tau)(\text{mg})$ is the amount of anesthetic drug in the third compartment at the time of target change.

The new continuous infusion rate formula takes into account the amount of drug already present in the peripheral compartments and which will be redistributed.

In the case when the Cpt decreases, no additional bolus is necessary, however the infusion of anesthetic drug is halted until $NCpt$ is achieved on central compartment due to elimination of drug presented on the PK model. Once this $NCpt$ is achieved the continuous infusion restarts according to Equation (9) where τ is the time $NCpt$ is achieved on central compartment through the elimination process.

2.4. Clinical Protocol

The parameters of PD model in this paper are determined by clinical data obtained from 42 elderly patients who had major vascular surgery of the lower limb. The information of these patients is summarized as follows (in the format of "mean ± standard deviation"): 35 males and 7 females, age 72 ± 8 years old, height 169 ± 10 cm and weight 73 ± 14 kg.

The process of collecting data from these patients is briefly described here. Before induction, all sensors were settled and a radial arterial line was inserted to receive baseline readings on all clinical monitors. Two Alaris PK syringe pumps were employed to implement the total intravenous anesthesia (TIVA) with propofol [27] and re-

mifentanil [28] TCI. To begin with, Cpt_{remi} was set to be 3 ng/ml. Once Ce_{remi} reached 2 ng/ml, Cpt_{remi} was altered to 2 ng/ml and Cpt_{prop} was started with 3 mg/ml. From then on, Cpt_{remi} and Cpt_{prop} were adjusted by anesthetists according to readings of different clinical apparatuses to maintain the BIS value within [40,60] for an appropriate DOA. Among these readings, cardiac output (CO) was monitored and stabilized at or near pre-induction level by anesthetists through all periods of the surgery.

3. PID CONTROLLERS

In this section, 3 types of PID controllers, namely linear PID controller, T1 fuzzy PID controller and IT2 fuzzy PID controller, are introduced. These 3 types of PID controllers are employed to regulate the output BIS index of the anesthesia model.

3.1. Linear PID controller

A linear PID controller is shown in Figure 3, which consists of 3 elements, namely proportional, integral and derivative blocks. The output of the linear PID controller in discrete time is given as

$$u(t_k)=K_P e(t_k)+K_I\sum_{i=0}^{k}e(t_i)\Delta t_k+K_D\frac{e(t_k)-e(t_{k-1})}{\Delta t_k}, \tag{10}$$

where $t_k=k\Delta t_k$, $k=0,1,2,\cdots$, is the sampling time; Δt_k is the interval of sampling time; $u(t_k)\in R$ is the output; $K_I\in R$, $K_P\in R$, and $K_D\in R$ are constant proportional, integral and derivative gains, respectively, to be determined.

3.2. Type-1 Fuzzy PID Controller

In view of the linear PID controller [33] , as the proportional, integral and derivative gains are constant, it is not able to handle well a highly nonlinear system. It motivates the use of T1 fuzzy PID controller [34] [35] of which the gains are changing according to the operating domains. By applying different sets of gains in different operating domains, more appropriate PID controller is employed to deal with the nonlinear system resulting in an improvement of control performance.

A T1 fuzzy PID control system is shown in Figure 4, which consists of a T1 fuzzy PID controller and a patient's model (detailed in Figure 1) connected in a closed loop. Unlike the linear PID controller having a set of constant gains, the T1 fuzzy PID controller has a fuzzy inference system providing a set of feedback gains through a reasoning process according to the operating condition.

The behavior of the fuzzy inference system is governed by a set of fuzzy rules of the following format:

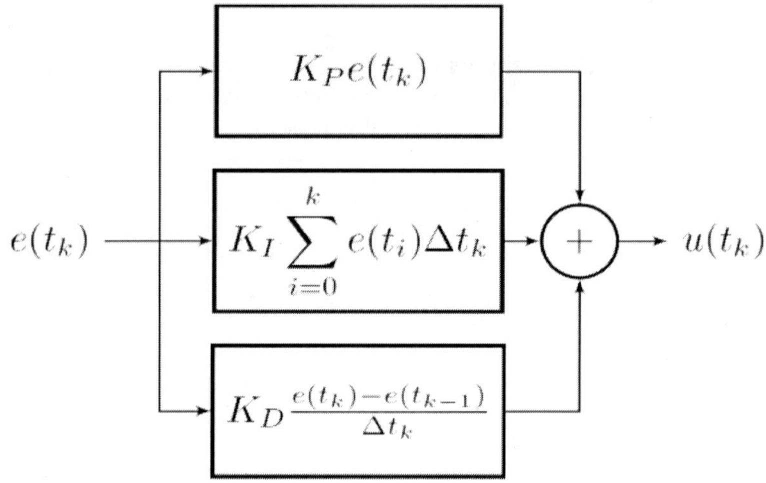

Figure 3. A block diagram of linear PID controller.

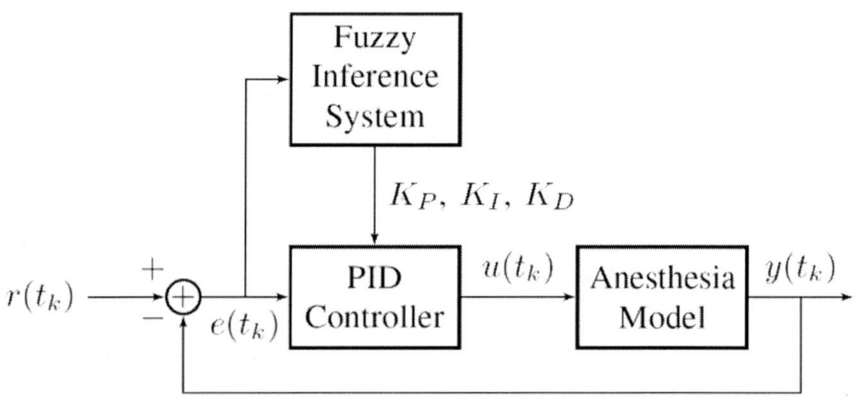

Figure 4. A block diagram of fuzzy PID control system.

$$\text{Rule } i: \quad \text{IF } x_1(t) \text{ is } M_1^i \text{ AND } \cdots \text{ AND } x_\Psi(t) \text{ is } M_\Psi^i$$
$$\text{THEN } y(\boldsymbol{x}(t)) = y_i, \tag{11}$$

where $x_j(t), j = 1,2,\cdots,\Psi,$ is a linguistic variable; $M_j^i, j = 1,2,\cdots,\Psi,$ is the fuzzy term corresponding to the linguistic variable x_j in the ith rule; Ψ is a positive integer; $y(\boldsymbol{x}(t))$ is the output of the fuzzy inference system; and y_i is the singleton membership function corresponding to the ith rule. The inferred output is gi- ven as

$$y(\boldsymbol{x}(t)) = \sum_{i=1}^{p} w_i(\boldsymbol{x}(t))\, y_i, \tag{12}$$

where

$$\boldsymbol{x}(t) = [x_1(t) \quad x_2(t) \quad \cdots \quad x_\Psi(t)],$$

$$\sum_{i=1}^{p} w_i(\boldsymbol{x}(t)) = 1, \tag{13}$$

$$w_i(\boldsymbol{x}(t)) \geq 0 \quad \forall i, \tag{14}$$

$$w_i(\boldsymbol{x}(t)) = \frac{\prod_{l=1}^{\Psi} \mu_{M_l^i}(\boldsymbol{x}(t))}{\sum_{k=1}^{p} \prod_{l=1}^{\Psi} \mu_{M_l^k}(\boldsymbol{x}(t))} \quad \forall i, \tag{15}$$

where $p > 0$ denotes the number of rules; $w_i(x(t))$ is the normalized grade of membership;

$\mu_{M_\alpha^i}(x(t)),\ \alpha = 1, 2, \cdots, \Psi,$ is the grade of membership corresponding to the fuzzy term M_α^i.

The output of the fuzzy inference system Equation (12) is employed to replace the gains of the linear PID controller turning it to become a T1 fuzzy PID controller. More precisely, 3 fuzzy inference systems are required to implement a T1 fuzzy PID controller. The outputs of the 3 fuzzy inference systems will be employed as K_P, K_I and K_D. Consequently, the PID gains are no longer constant but dependent on the operating condition characterized by the membership functions.

3.3. Interval Type-2 Fuzzy PID Controller

Type-2 (T2) fuzzy sets [36] -[39] demonstrate a superior characteristic handing uncertainties compared with the T1 fuzzy sets. Uncertainties are captured by the lower and upper membership functions which form the footprint of uncertainty (FOU). A T2 fuzzy inference system can be considered as a set of infinite number of T1 fuzzy inference systems. Consequently, T2 fuzzy inference system is able to outperform the T1 fuzzy inference system in terms of reasoning and generalization capability. In general, the defuzzification process for general T2 fuzzy sets is computational demanding. By using IT2 fuzzy sets [40] , the computational demand can be significantly reduced.

By employing interval fuzzy sets for the T1 fuzzy inference system Equation (12), it becomes an IT2 fuzzy inference system. The behavior of an IT2 fuzzy inference system is described by a set of rules of the following format:

$$\begin{aligned}\text{Rule } i\text{:}\quad & \text{IF } x_1(t) \text{ is } \tilde{M}_1^i \text{ AND } \cdots \text{ AND } x_\Psi(t) \text{ is } \tilde{M}_\Psi^i \\ & \text{THEN } y(x(t)) = \tilde{y}_i,\end{aligned} \tag{16}$$

$$W_i(x(t)) = \left[\underline{w}_i(x(t)), \quad \overline{w}_i(x(t))\right], \tag{17}$$

$$\tilde{y}_i = \left[\underline{y}_i, \quad \overline{y}_i\right], \ i = 1, 2, \cdots, p, \tag{18}$$

Where

$$\underline{w}_i(x(t)) = \prod_{\alpha=1}^{\Psi} \underline{\mu}_{\tilde{M}_\alpha^i}(x(t)) \geq 0, \tag{19}$$

$$\overline{w}_i(x(t)) = \prod_{\alpha=1}^{\Psi} \overline{\mu}_{\tilde{M}_\alpha^i}(x(t)) \geq 0, \tag{20}$$

$$\overline{\mu}_{\tilde{M}_\alpha^i}(x(t)) \geq \underline{\mu}_{\tilde{M}_\alpha^i}(x(t)) \geq 0, \tag{21}$$

$$\overline{w}_i(x(t)) \geq \underline{w}_i(x(t)) \geq 0, \tag{22}$$

$$\overline{y}_i \geq \underline{y}_i \geq 0, \ \forall i, \tag{23}$$

in which $\underline{w}_i(x(t))$, $\overline{w}_i(x(t))$, $\underline{\mu}_{\tilde{M}_\alpha^i}(x(t))$ and $\overline{\mu}_{\tilde{M}_\alpha^i}(x(t))$ denote the lower grade of membership, upper grade of membership, lower membership function and upper membership function, respectively. Using the center-of- set type reducer, the inferred output of the IT2 fuzzy inference system is given as

$$y(x(t)) = \bigcup_{\tilde{w}_i(x(t)) \in W_i(x(t)), y_i \in \tilde{y}_i} \frac{\sum_{i=1}^{p} \tilde{w}_i(x(t)) y_i}{\sum_{i=1}^{p} \tilde{w}_i(x(t))} = [y_l, \ y_r], \tag{24}$$

where y_l and y_r can be obtained using the Karnik-Mendel (KM) algorithms [41] . The final defuzzified output is given by

$$y(x(t)) = \frac{y_l + y_r}{2}. \tag{25}$$

The IT2 fuzzy PID control system [42] can be represented by Figure 4 as well of which the PID gains are given by the outputs of 3 IT2 fuzzy inference systems.

4. SIMULATION DESIGN

In this section, the simulation environment is presented. First, we design the training profiles and testing profiles which all controllers need to deal with. The purpose of selected target profiles is first described. Then different control strategies are provided for comparison. For fuzzy control strategies, the approach to determine fuzzy rules is offered. Finally, the procedure of using GA for optimization is described, and a performance index is designed as the cost function for optimization.

4.1. Target Profiles

A training profile as shown in Figure 5, which is a series of BIS values required for an appropriate DOA, is employed for the training of the PID control strategies using GA. It defines the target BIS values in different periods that a PID control strategy has to achieve. After the training, a testing profile as shown in Figure 6 is employed to verify if the trained PID control strategies are able to control the BIS value to reach some targets subject to different operating conditions.

Clinically, the induction process starts with $\mathrm{BIS}(t_k)=98$. As a result, the initial condition for the simulation is $\mathrm{BIS}(0)=98$ and the target profile begins by $\mathrm{BIS}_t(t_k)=98$, where $\mathrm{BIS}_t(t_k)$ is the target BIS at sampling time t_k. At $t_k=120$ seconds, the induction is executed and $\mathrm{BIS}_t(t_k)=50$. Since the induction period is short compared with the maintenance period, there is not much difference during induction in terms of control performance between various control strategies. Thus, we use two linear PID controllers for two drugs to drive the BIS from 98 to 50. By trial and error, the PID gains are predefined, and it is guaranteed that $\mathrm{BIS}(1000)$ is around 50 and the maintenance process starts from $t_k=1000$ seconds.

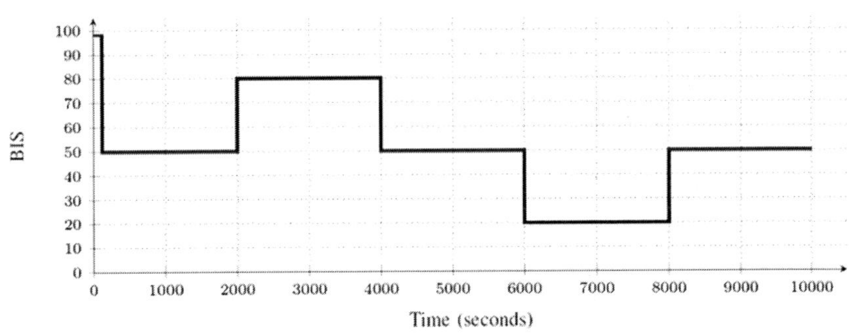

Figure 5. Training profile.

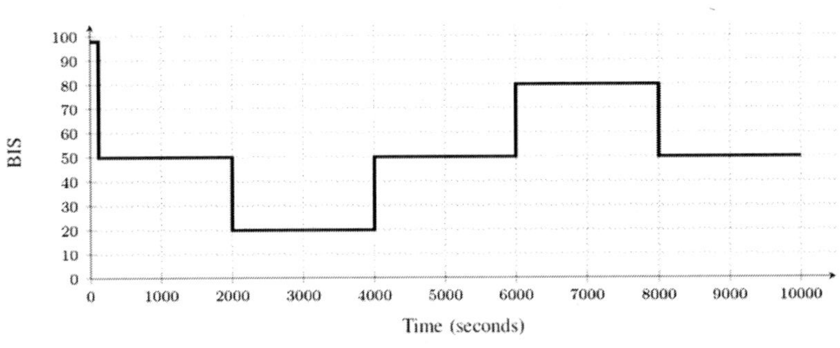

Figure 6. Testing profile.

At $t_k = 1000$ seconds, designed controllers replace the predefined controller to achieve the maintenance process. For a suitable anesthesia, it is required that $\mathrm{BIS}_t(t_k) = 50$ and an acceptable range is $\mathrm{BIS} \in [40,60]$. Hence, the control objective is to maintain the BIS value within [40,60]. In order to make the control difficult enough to differentiate various control strategies, we add more targets into the profiles such as $\mathrm{BIS}_t(t_k) = 80,\ t_k \in [2000,4000)$ and $\mathrm{BIS}_t(t_k) = 20,\ t_k \in [6000,8000)$ in Figure 5.

As for the recovery process, since it is not allowed take the drug out of humans, the cease of controllers is the only and fastest option. These will not exist overshoot either because the target BIS is the maximum value. Therefore, all controllers will perform the same at this stage. It is not necessary to design controllers for the recovery process which is thus not included in the target profiles. It is noted that although the induction process is not under comparison either, we cannot ignore it. The reason is that it is difficult to find all model parameters describing the state of $\mathrm{BIS}_t(t_k) = 50$. On the other hand, the parameters for $\mathrm{BIS}_t(t_k) = 98$ are known and it is easy to start from this initial condition.

4.2. Control Strategies

In this paper, we aim to compare different control strategies to achieve a better control performance. All six control strategies are listed in Table 1. These six cases can be separated into two groups: two-controller Cases 1 to 3 and one-controller Cases 4 to 6. In two-controller cases, we control each drug by an independent PID controller offering control outputs $u_p(t_k)$ and $u_r(t_k)$. In one-controller cases, however, the control output of the PID controller is divided into two control outputs by two scaling factors α_r and α_p. Apparently, parameters of these two divided controllers are constrained by the ratio of these two factors. Theoretically, this constraint leads to conservativeness and make the performance worse than the ones of two-controller cases. Nevertheless, one-controller cases have less number of parameters to be determined resulting in lower computational demand, faster convergence in the training process and lower implementation cost for the PID control strategies.

Table 1. Six cases of PID control strategies.

Case	Description
1	Two linear PID controllers
2	Two T1 fuzzy PID controllers
3	Two IT2 fuzzy PID controllers
4	One linear PID controller with scaling factors
5	One T1 fuzzy PID controller with T1 fuzzy scaling factors
6	One IT2 fuzzy PID controller with IT2 fuzzy scaling factors

In each of these two groups, we have the following control strategies: linear PID controllers, T1 fuzzy PID controllers, and IT2 fuzzy PID controllers. For fuzzy PID controllers, the block diagram of BIS index regulation using two controllers and one controller with scaling factors are shown inFigure 7 and Figure 8, respectively. Referring to the figures, $r(t)$ denotes the desired BIS target value, $y(t_k)$ is the output BIS value from theanesthesia model, $e(t_k)$ denotes the difference between the desired target and actual BIS value at time t_k, i.e. $e(t_k)=r(t_k)-y(t_k)$. The PID gains $(K_{P_1}, K_{I_1}, K_{D_1}, K_{P_2}, K_{I_2} \text{ and } K_{D_2})$ and scaling factors $(\alpha_r \text{ and } \alpha_p)$ are decided by the fuzzy inference systems. The PID controller will generate the control signals $u_p(t_k)$ and $u_r(t_k)$ for BIS regulation where $u_p(t_k)$ and $u_r(t_k)$ are Cpt_{prop} and Cpt_{remi} in Figure 1.

4.3. Fuzzy Rules

T1 and IT2 fuzzy PID controllers are discussed in Subsections 3.2 and 3.3, respectively. While the number of fuzzy rules and the shape of membership function are predefined, the parameters of input and output membership functions are optimized by GA.

In this paper, the shape of membership functions is defined as triangular shape for simplicity. The triangular IT2 membership functions are shown in Figure 9. For each input IT2 membership functions, the lower and upper membership functions are characterized by seven points p_1 to p_7 which are to be optimized by GA. As for T1 fuzzy PID controller, each input membership function is characterized by three points p_1 to p_3 which are to be optimized by GA. For both T1 and IT2 fuzzy PID controllers, their output membership functions are T1 and IT2 singleton membership functions whose values are to be determined by GA.

In fact, from the determination of membership functions and fuzzy rules, we can find the relation between these control strategies: solutions from linear PID controllers can be implemented by T1 fuzzy PID controllers, and solutions from T1 fuzzy PID controllers can be implemented by IT2 fuzzy PID controllers. In other words, linear PID controller is a subset of T1 fuzzy PID controller, and T1 fuzzy PID controller is a subset of IT2 fuzzy PID controller. By adding some constraints on the membership functions and consequents in fuzzy rules, IT2 fuzzy PID controllers can be reduced to T1 fuzzy PID controllers, and T1 fuzzy PID controllers can be reduced to linear PID controllers.

Three rules are employed for each fuzzy inference system. The premise of each rule takes only one linguistic variable, i.e., $\Delta\mathrm{BIS}(t_k) = \mathrm{BIS}_t(t_k) - \mathrm{BIS}(t_k)$ where $\mathrm{BIS}_t(t_k)$ is the target BIS value at time t_k and $\mathrm{BIS}(t_k)$ is the actual BIS value. Three linguistic terms, namely negative (N), zero (Z), and positive (P), are employed to characterize the premise variable $\Delta\mathrm{BIS}(t_k)$. More rules can be used to partition the universe of discourse for a better result. However, it will lead to slower convergence of training and more computational burden. Additionally, the rate of change of $\Delta\mathrm{BIS}(t_k)$ can also be treated as another linguistic variable together with $\Delta\mathrm{BIS}(t_k)$. Likewise, it will cause more fuzzy rules and parameters which increase the computational burden and defers the convergence.

From the above discussion, triangular membership functions and three rules are employed for both T1 and IT2 fuzzy PID controllers. In the GA optimization, the triangular shape and sequence of membership functions of three rules should be guaranteed. For the sequence, specifically, the same point in membership functions corresponding to linguistic terms N, Z and P should be in ascending order (except p_7). For example, $N_{p2} < Z_{p2} < P_{p2}$, where N_{p2}, Z_{p2}, and P_{p2} are points p_2 for membership functions corresponding to linguistic terms N, Z and P, respectively.

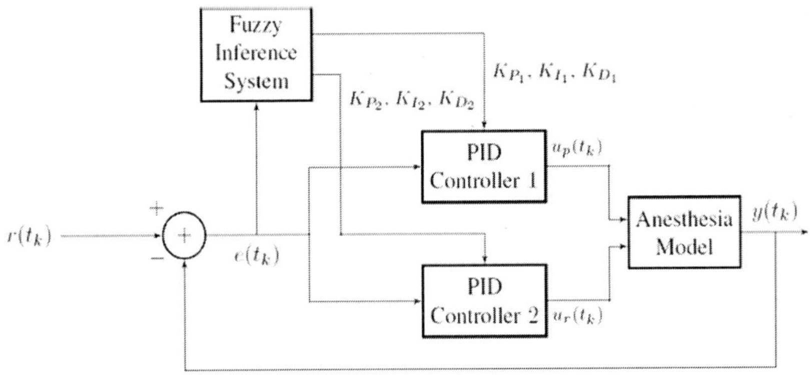

Figure 7. BIS index regulation using two fuzzy PID controllers.

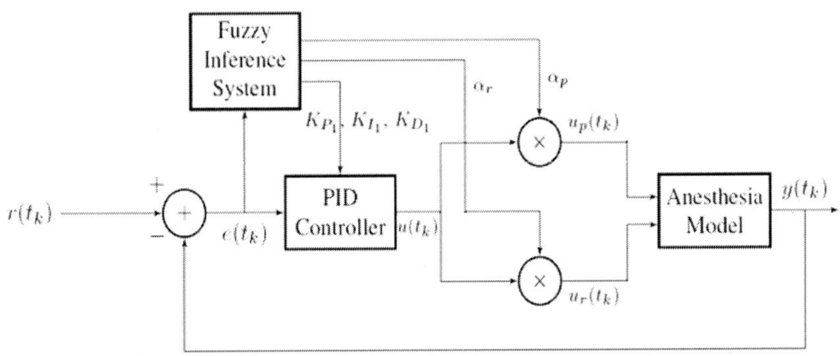

Figure 8. BIS index regulation using one fuzzy PID controller with scaling factors.

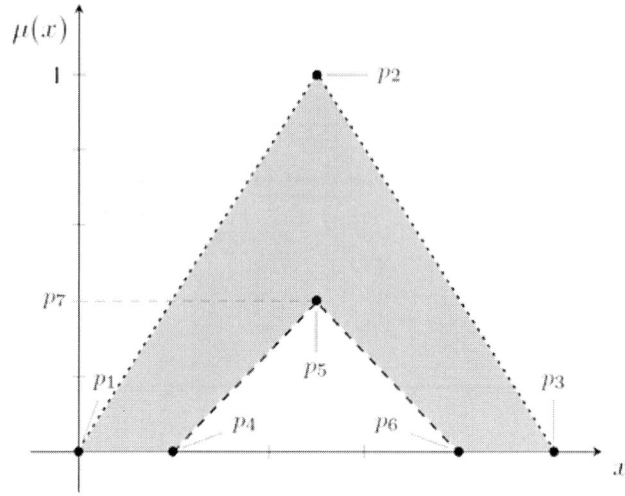

Figure 9. An example of IT2 membership functions. Dashed line: lower membership function. Dotted line: Upper member- ship function. Gray area: footprint of uncertainty.

Another condition needed to be ensured is that at least one rule is fired for $\Delta\text{BIS}(t_k) \in [-100,100]$. Between two adjacent membership functions, as long as they intersect each other, there exists one rule to be fired. On the left-hand side of membership function N and the right-hand side of membership function P, there are two approaches to ensure this condition. One is defining $N_{p2} = 100$ and $P_{p2} = 100$, and another is defining $N_{p1} = N_{p4} = -\infty$ and $P_{p3} = P_{p6} = +\infty$. It can be found that solutions from the first approach can be implemented by the second approach. In other words, the second approach is more general than the first approach. Hence, we adopt the second method. For that reason, rules N and P are two shoulder-shape membership functions which can be treated as a special case of triangular membership functions. In this approach, it is guaranteed that at least one rule is fired for $\Delta\text{BIS}(t_k) \in [-\infty,+\infty]$, although only $[-100,100]$ is concerned practically.

Note that under this method the equalities $N_{p1} = N_{p4} = N_{p2}$ and $P_{p3} = P_{p6} = P_{p2}$ are temporarily employed in the optimization.

4.4. Parameters Optimization

The PID gains, scaling factors and membership functions are optimized by GA subject to a cost function reflecting the control performance. Due to the disparity of each training, GA is run for 10 times for each case of PID control strategies as shown in Table 1. The best set of parameters for each control strategy for each run of GA is recorded for further comparison, analysis and practical application. Statistical information including the worst, mean and best costs and the standard deviation for the 10 runs are collected. Among the 10 runs for each PID control strategy, the best set of parameters given by the best cost is used to implement the corresponding PID controller. During the optimization, the lower and upper bounds (LB and UB) of PID gains and scaling factors are listed in Table 2, which are determined by trial and error for good control performance.

Due to a large number of variables and the highly nonlinear cost function (defined in the following subsection 4.5), especially for fuzzy PID controllers, GA may not be able to reach the global optimal solution. To speed up the training process, we define the initial population for fuzzy PID controllers based on our knowledge on the PID control strategy. It is known that linear PID controller is a subset of T1 fuzzy PID controller, and T1 fuzzy PID controller is a subset of IT2 fuzzy PID controller. It is thus reasonable that the best set of PID gains for linear PID controllers obtained by GA is employed as the initial PID gains for T1 fuzzy PID controllers for all rules. As a result, initially, the T1 fuzzy controller is equivalent to a linear PID controller. Similarly, the best set of PID gains and membership function parameters obtained for T1 fuzzy PID controller is employed as the initial set of PID gains and membership function parameters for IT2 fuzzy PID controller. That is to say, we utilize the "knowledge" on the PID control strategy such that GA starts with a considerable well initial condition. In this way, although there is still the same number of variables to be trained, GA only needs to utilize the additional parameters and new structures (fuzzy rules and membership functions) provided by fuzzy PID controllers in order to obtain a better cost value.

4.5. Performance Index

The performance index is used to judge whether the control objective is achieved and show the merits of various control strategies. Different performance indices can be selected such as settling time, overshoot, steady error, mean absolute error (MAE), and mean square error (MSE). For the GA optimization, a well-defined performance index is required as the cost function. All parameters of controllers are optimized according to the cost value provided by the cost function. In this paper, we present the following index J mainly based on MAE:

$$J = \frac{1}{n+1}\sum_{k=0}^{n}\left(\lambda_1\left|\Delta\mathrm{BIS}(t_k)\right| + \lambda_2\left|\frac{\Delta\mathrm{BIS}(t_k)-\Delta\mathrm{BIS}(t_{k-1})}{T}\right| + \lambda_3\left|\frac{Cpt_{prop}(t_k)-Cpt_{prop}(t_{k-1})}{T}\right|\right.$$
$$\left. + \lambda_4\left|\frac{Cpt_{remi}(t_k)-Cpt_{remi}(t_{k-1})}{T}\right|\right) + \lambda_5\left|\frac{\sum_{k=0}^{n}\left(Cpt_{prop}(t_k)\right)}{\sum_{k=0}^{n}\left(Cpt_{prop}(t_k)+Cpt_{remi}(t_k)\right)} - 0.5\right|, \quad (26)$$

where $t_k = kT,\ k = 0,1,2,\cdots,n,$ is the sampling time, and n is a positive integer; T is the interval of sampling time; $\Delta\mathrm{BIS}(t_K) = \mathrm{BIS}_t(t_k) - \mathrm{BIS}(t_k)$, $\mathrm{BIS}_t(t_k)$ is the target BIS value at time t_k, and $\Delta\mathrm{BIS}(t_{-1}) = 0$; $Cpt_{prop}(t_k)$ and $Cpt_{remi}(t_k)$ are plasmatic concentration targets of propofol and remifentanil, respectively, and $Cpt_{prop}(t_{-1}) = Cpt_{prop}(t_0)$, $Cpt_{remi}(t_{-1}) = Cpt_{remi}(t_0)$; $\lambda_1, \lambda_2, \cdots, \lambda_5$ are predefined weights.

In the cost function Equation (26), the first term is MAE which aims at minimizing the difference between the current BIS value and its target. Unlike settling time, overshoot and steady error, MAE and MSE record the error information through all simulation periods which reflect more comprehensive properties of performance. The reason we choose MAE instead of MSE is that MSE amplifies the effect of large errors and then the optimization tries to minimize large errors which results in oscillation. On the other hand, MAE keeps the original weights on large and small errors leading to a mild and smooth response.

Table 2. Lower and upper bounds of parameters.

K_{P1}, K_{P2}		K_{I1}, K_{I2}		K_{D1}, K_{D2}		α_p		α_r	
LB	**UB**	**LB**	**UB**	**LB**	**UB**	**LB**	**UB**	**LB**	**UB**
−10	0	−0.1	0	−10	0	0	1	0	1

The three terms after the first term of (26) are the rate of change of $\Delta \text{BIS}(t_k)$, $\text{BIS}(t_k)$, $Cpt_{prop}(t_k)$ and $Cpt_{remi}(t_k)$, respectively, which are designed to reduce the oscillation for a gentle progress. The last term is to reduce the difference of the concentration of two drugs because they are equally important for different purposes during the anesthesia. Serious bias to either of these two drugs will not provide an adequate anesthesia.

5. SIMULATION RESULTS

In this section, we implement the simulation on the control of anesthesia using the anesthesia model in Figure 1. Different control strategies in Table 1 are applied to regulate the BIS using the training profiles shown in Figure 5 for training and their control performance are verified by testing profiles shown in Figure 6. PID gains, parameters of scaling factors and membership functions are optimized by GA according to the cost function Equation (26). Comparisons of performance are made between the six cases. It should be noted that the simulation was carried out in discrete time. The PK model in Equation (1) was discretized by zero-order-hold (ZOH) assumption. The TCI system implemented has also been adjusted to perform under discrete PK model.

A real-coded GA available from Matlab global optimization toolbox is employed for the training. The control parameters of GA are listed in Table 3.

We define the parameters in Equation (26) as follows: $T = 10$ $n = 1000$, seconds, $\lambda_2 = \lambda_3 = \lambda_4 = 20$ $\lambda_1 = \lambda_5 = 1$,. It is worth to mention that although the weight λ_1 is not the largest, the MAE is still the main contribution for the total cost. The bounds of parameters in Table 2 are adopted for the training process of GA. By running GA 10 times for each case in Table 1, we obtain the statistical information of the cost J as shown in Table 4.

Referring to Table 4, comparing with Cases 1 to 3, Case 3 offers the best cost and Case 2 comes to the second which coincides with the theory. The same rank can be found in Cases 4 to 6. Comparing Cases 1 and 4, two- controller case is better than one-controller case, which can be also proved by comparing between Cases 2 and 5, and Cases 3 and 6. The reason is that one-controller case has the constraint on the ratio of control signal between two drugs while two-controller case does not have such constraint. One-controller case is a subset of two-con- troller case. Although one-controller case performs worse, it has less number of parameters to be determined and thus reduces the computational demand and implementation cost of PID control strategy.

In spite of the "Best" cost, the "Std" is descending from linear PID controllers to IT2 fuzzy PID controller ex- cept Case 4. It provides the information that the convergence gets easier for fuzzy PID controllers. The reason is that we use the best set of parameters of previous cases as the initial populations of following training. It also in- dicates that the improvement becomes more and more difficult, and thus 10 times running offer similar results. The "Std" for two-controller cases is larger than the corresponding one-controller cases, which implies that the convergence for two-controller cases is harder due to more variables to be trained.

The best sets of parameters and membership functions are shown in Tables 5-14. Corresponding membership functions are exhibited in Figures 10-13. From the obtained membership functions, it can be summarized that points $|N_{p2}|$ and $|P_{p2}|$ are preferred to be around 20 related to the training profiles we define in Figure 5. Considering that the only linguistic variable is $\Delta\text{BIS}(t_k)$ and the maximum change of $\text{BIS}_r(t_k)$ is ± 30 in the training profile, therefore, it is reasonable to choose $\Delta\text{BIS}(t_k) \in [-30, 30]$ as an estimated domain where the PID controllers work suggesting that the fuzzy blending should start from around ± 20. For the IT2 member- ship functions of IT2 fuzzy PID controllers in Figure 11 and Figure 13 after training, it was found that the lower and upper membership functions are very close to each other such as membership functions of N and Z in Figure 11. The values of points $(p_1 \text{ to } p_7)$ associated with the IT2 membership functions are shown in Table 9 and Table 14.

By applying the best sets of parameters to the training profile of BIS, the time responses of BIS and corresponding drug concentration information are shown in Figures 14-19 for Cases 1 to 6, respectively. By applying the best sets of parameters to the testing profile of BIS, the ones for the testing profile are exhibited in Figures 20-25. The green line is drawn by $\mathrm{BIS}(t_k) = \mathrm{BIS}_t(t_k) \pm 2$. From the practical clinical point of view, the acceptable range of BIS is $[\mathrm{BIS}_t(t_k) - 10, \mathrm{BIS}_t(t_k) + 10]$. Therefore, if BIS is stabilized within the region bounded by the lower and upper green lines, the performance will be well acceptable.

Table 3. Control parameters of GA.

Parameter	Value
Number of iterations	200
Size of population	100
Selection	Stochastic uniform selection function.
Elitism	Elitism is implemented. The best two chromosomes are guaranteed to survive to the next population.
Crossover	Scattered crossover.
Crossover fraction	0.8
Mutation	Gaussian mutation.
Stopping criterion	It stops if the weighted average relative change in the best fitness function value over 100 generations is less than or equal to 10^{-6}.

Table 4. The cost J from running GA 10 times.

Case	Best	Worst	Std	Mean
1	4.0546	6.0544	0.6915	4.7190
2	3.8913	4.0479	0.0439	3.9273
3	3.8606	3.8899	0.0091	3.8825
4	4.0585	4.0594	0.0003	4.0587
5	3.9544	3.9780	0.0073	3.9696
6	3.9264	3.9437	0.0060	3.9380

Table 5. Best set of parameters for two PID controller.

K_{P1}	K_{I1}	K_{D1}
−0.0531	−0.0004	−0.0028
K_{P2}	K_{I2}	K_{D2}
−0.2203	−0.0012	−0.0910

Table 6. Best set of parameters for two T1 fuzzy PID controllers.

Rule	K_{P1}	K_{I1}	K_{D1}
N	−0.0531	−0.0002	−0.5027
Z	−1.8031	−0.0004	−4.0368
P	−2.5531	−0.0004	−0.0028
	K_{P2}	K_{I2}	K_{D2}
N	−0.2203	−0.0012	−0.0910
Z	−0.2203	−0.0012	−1.5909
P	−7.5162	−0.0012	−0.5910

Table 7. Best set of parameters for two T1 fuzzy PID controllers.

Rule	p_1	p_2	p_3
N	$-\infty$	−19.0370	7.0855
Z	−7.1950	5.7480	13.1011
P	−1.1593	19.0370	$+\infty$

Table 8. Best set of parameters for two IT2 fuzzy PID controllers.

Rule	K_{P1}	K_{I1}
N	$[-0.0531, -0.0530]$	$[-0.0002, -0.0002]$
Z	$[-1.8031, -1.8022]$	$[-0.0004, -0.0004]$
P	$[-2.5531, -2.5481]$	$[-0.0020, -0.0020]$
	K_{D1}	K_{P2}
N	$[-0.5027, -0.4399]$	$[-0.2203, -0.2203]$
Z	$[-6.5368, -6.5352]$	$[-0.2203, -0.2203]$
P	$[-0.0013, -0.0013]$	$[-7.5162, -3.7581]$
	K_{I2}	K_{D2}
N	$[-0.0012, -0.0012]$	$[-0.0910, -0.0451]$
Z	$[-0.0012, -0.0012]$	$[-1.5909, -1.5870]$
P	$[-0.0012, -0.0012]$	$[-0.5910, -0.5310]$

Table 9. Best set of membership functions for two IT2 fuzzy PID controllers.

Rule	p_1	p_2	p_3	p_4
N	$-\infty$	−19.0371	7.0975	$-\infty$
Z	−7.2069	5.7480	13.4781	−7.1951
P	−1.4565	19.0371	$+\infty$	−1.1593
	p_5	p_6	p_7	
N	−19.0371	7.0856	0.9990	
Z	5.7480	13.1011	0.9990	
P	19.0371	$+\infty$	0.8135	

Table 10. Best set of parameters for one PID controller.

K_{P1}	K_{I1}	K_{D1}	α_r	α_e
−0.9267	−0.0061	−0.6068	0.0598	0.1869

Table 11. Best set of parameters for one T1 fuzzy PID controller with T1 fuzzy scaling factors.

Rule	K_{P1}	K_{I1}	K_{D1}	α_p	α_r
N	−0.8642	−0.0061	−0.5443	0.0665	0.2189
Z	−1.7068	−0.0054	−9.6913	0.0725	0.1869
P	−8.6575	−0.0061	−6.6119	0.0602	0.1947

Table 12. Best set of membership functions for one T1 fuzzy PID controller with T1 fuzzy scaling factors.

Rule	p_1	p_2	p_3
N	$-\infty$	−24.6107	−10.1748
Z	−11.4184	−6.9156	1.8562
P	−9.1078	24.6107	$+\infty$

Table 13. Best set of parameters for one IT2 fuzzy PID controller with IT2 fuzzy scaling factors.

Rule	K_{P1}	K_{I1}
N	**[−0.8642, −0.8372]**	[−0.0061, −0.0061]
Z	[−1.7068, −1.6635]	[−0.0032, −0.0031]
P	[−8.6575, −8.6575]	[−0.0061, −0.0059]
	K_{D1}	α_p
N	[−0.5443, −0.5273]	[0.0665, 0.0742]
Z	[−9.6913, −9.6393]	[0.1037, 0.1248]
P	[−7.2369, −7.2228]	[0.0563, 0.0594]
	α_r	
N	[0.2189, 0.2204]	
Z	[0.2162, 0.2407]	
P	[0.2045, 0.2728]	

Table 14. Best set of membership functions for one IT2 fuzzy PID controller with IT2 fuzzy scaling factors.

Rule	p_1	p_2	p_3	p_4
N	$-\infty$	−24.6105	−5.0849	$-\infty$
Z	−13.6412	−6.9156	2.2345	−13.6302
P	−9.1233	24.6105	$+\infty$	−9.1078
	p_5	p_6	p_7	
N	−24.6105	−7.0984	0.9990	
Z	−6.9156	1.8562	0.9990	
P	24.6105	$+\infty$	0.9990	

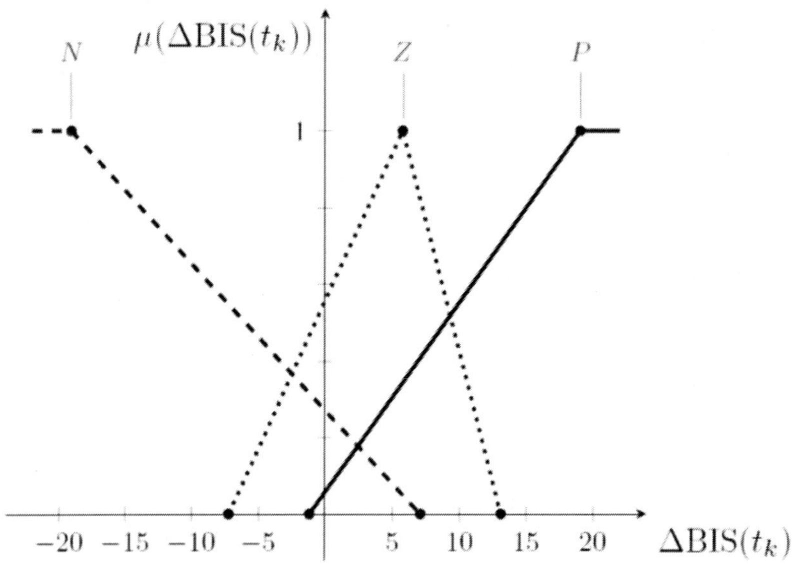

Figure 10. Membership functions for two T1 fuzzy PID con- trollers. Dashed line: membership function N. Dotted line: mem- bership function Z. Solid line: membership function P.

It can be seen from the figures that all cases of PID control strategies succeed in tracking the BIS targets and maintaining BIS values. In terms of control performance, while the linear PID controllers have less steady error, fuzzy PID controllers have less overshoot and settling time. Since there is no rigorous requirement on small steady error, fuzzy PID controllers are preferable that they are able to keep the BIS values within the regions bounded by the green lines with less overshoot and settling time. Both training and testing profiles demonstrate the same characteristics as discussed for the corresponding control strategies. Recalling a common controller with two linear PID controllers and predefined parameters is used to achieve the induction process for all cases, the proposed PID control strategies kick in at the start of the maintenance process. The switching point of induc- tion and maintenance process occurs at 1000 seconds which results in a small vibration.

The lower panels of Figures 20-25 show the behavior of drug concentration. It can be seen that both Cp and Ce of propofol and remifentanil are adjusted by controllers according to target profiles. More importantly, they are adjusted in advance to decrease the overshoot. This is mainly because all controllers are based on PID control strategy, whose abilities are inherited. In addition, the total amount of remifentanil is larger than propofol, which illustrates that the remifentanil is more favorable to the stabilization of BIS than the propofol. This might be due to the faster action time of remifentanil comparatively to propofol. Although the total amount of remifentanil is larger, the cost function takes the difference of the amount of two drugs into consideration such that serious bias to remifentanil is avoided. Even though the slightly bias towards remifentanil can be noticed overall the propofol and remifentanil Ce reached are within acceptable clinical boundaries even when the BIS target was set at 20, a value inferior to the safe interval of 40 to 60.

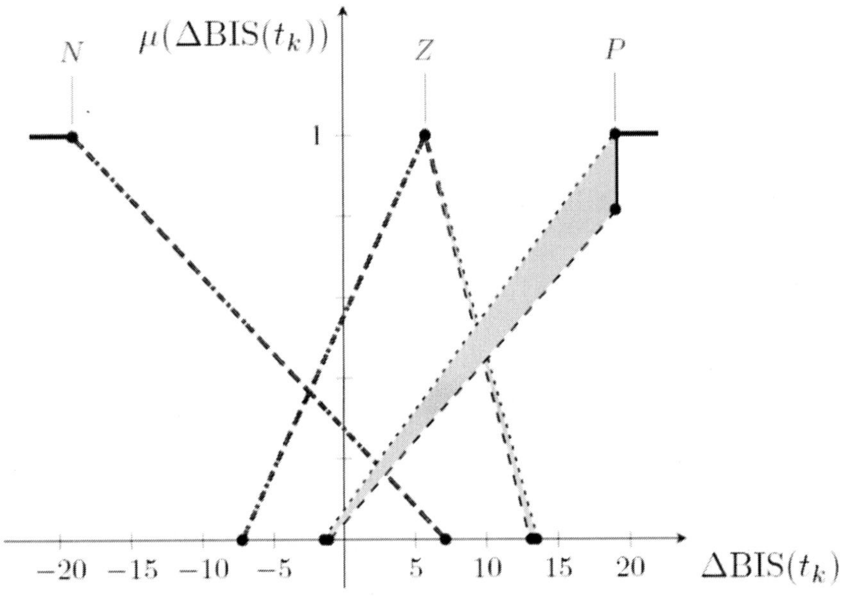

Figure 11. Membership functions for two IT2 fuzzy PID con- trollers. Dashed line: lower membership functions. Dotted line: upper membership functions. Solid line: the shoulder of membership functions.

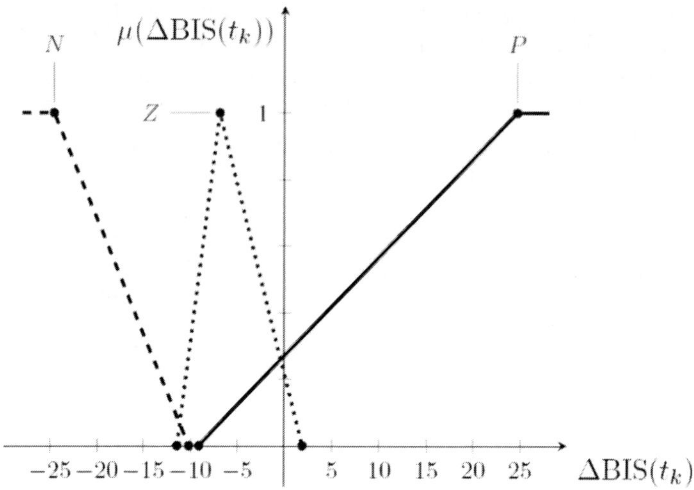

Figure 12. Membership functions for one T1 fuzzy PID con- troller with T1 fuzzy scaling factors. Dashed line: membership function N. Dotted line: membership function Z. Solid line: membership function P.

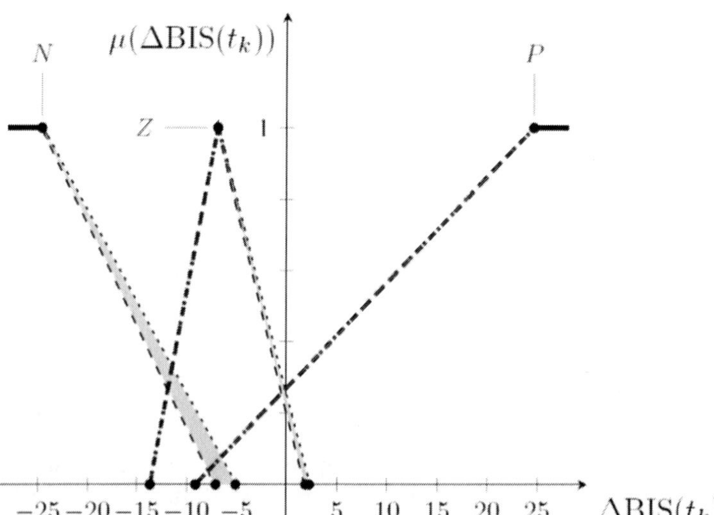

Figure 13. Membership functions for one IT2 fuzzy PID con- trollers with IT2 fuzzy scaling factors. Dashed line: lower membership functions. Dotted line: upper membership func- tions. Solid line: the shoulder of membership functions.

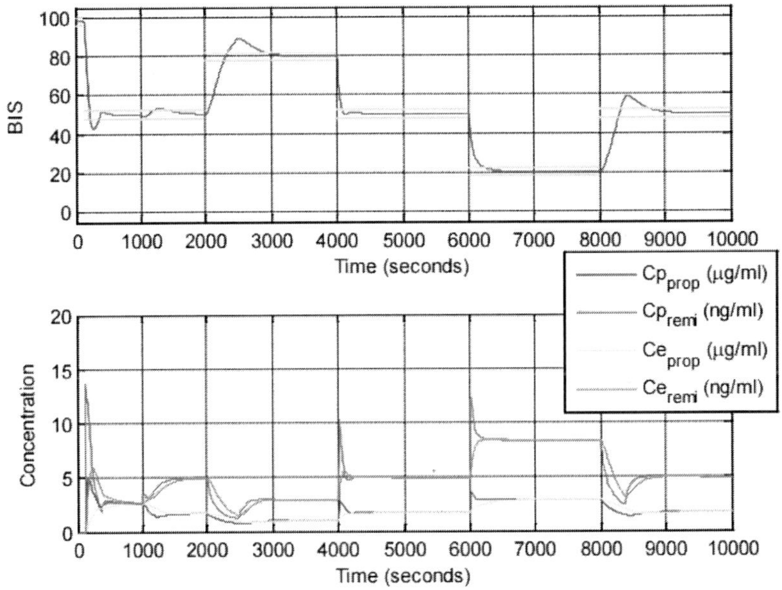

Figure 14. BIS and drug concentration for training profile by two PID controllers.

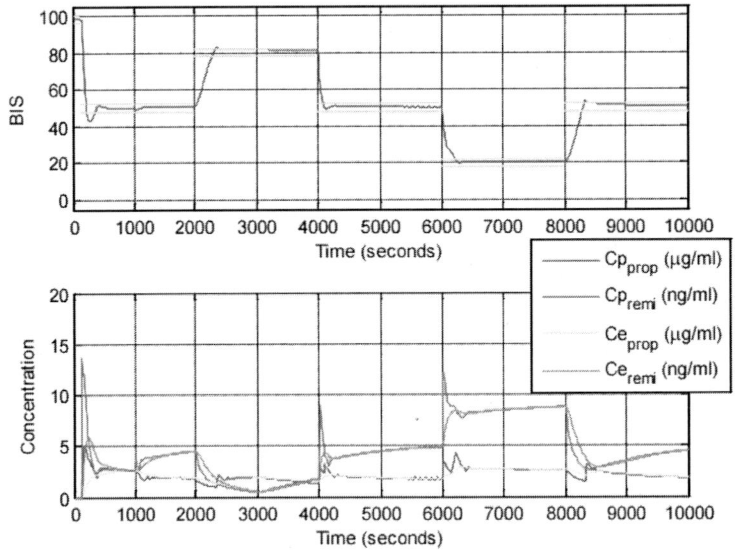

Figure 15. BIS and drug concentration for training profile by two T1 fuzzy PID controllers.

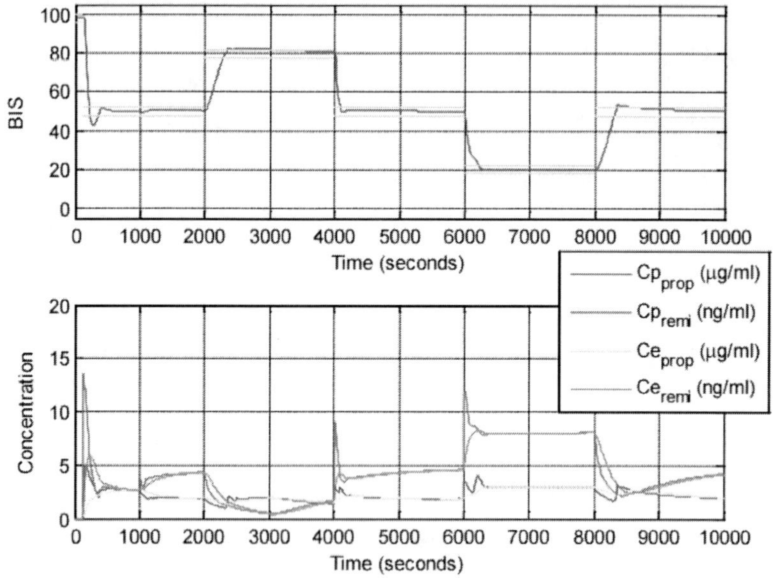

Figure 16. BIS and drug concentration for training profile by two IT2 fuzzy PID controllers.

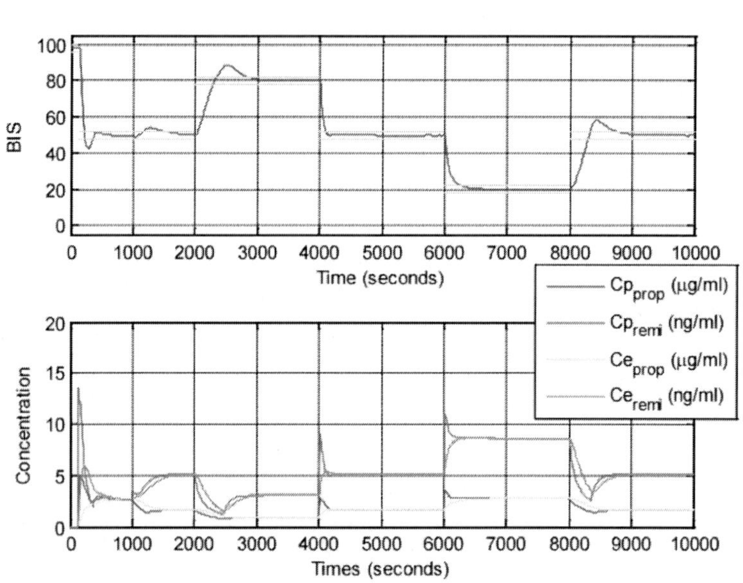

Figure 17. BIS and drug concentration for training profile by one PID controller with scaling factors.

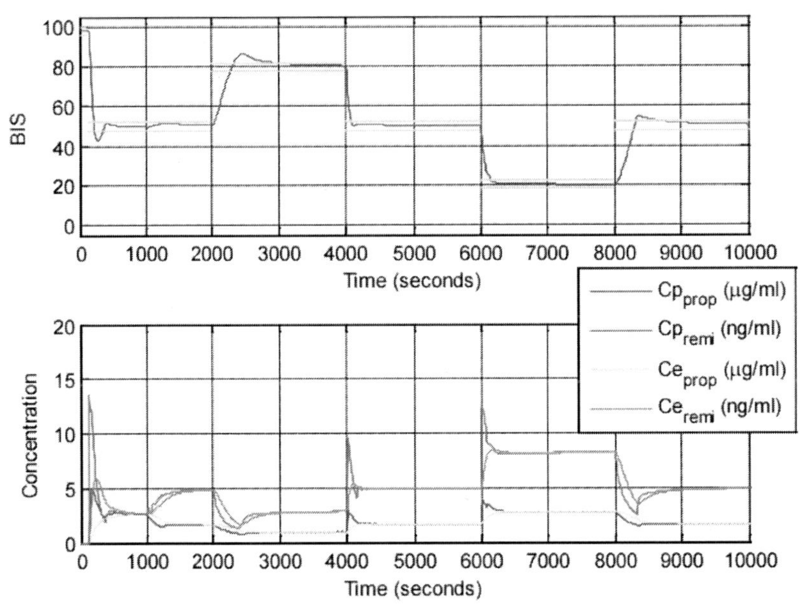

Figure 18. BIS and drug concentration for training profile by one T1 fuzzy PID controller with T1 fuzzy scaling factors.

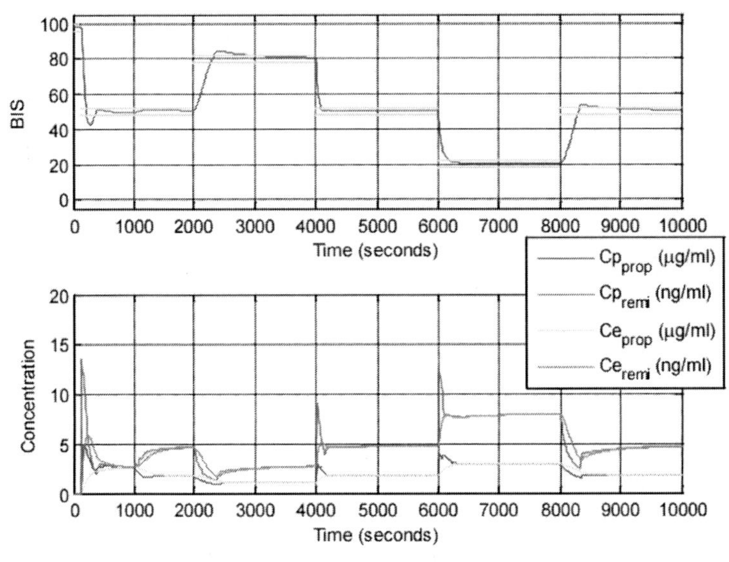

Figure 19. BIS and drug concentration for training profile by one IT2 fuzzy PID controller with IT2 fuzzy scaling factors.

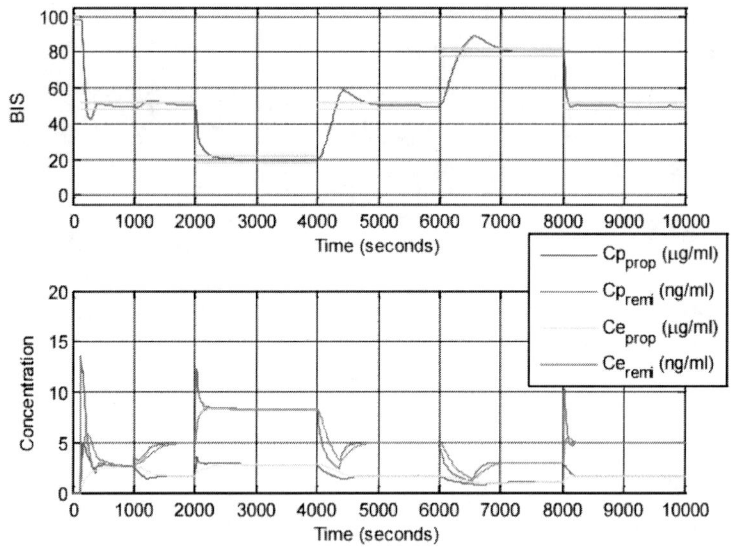

Figure 20. BIS and drug concentration for testing profile by two PID controllers.

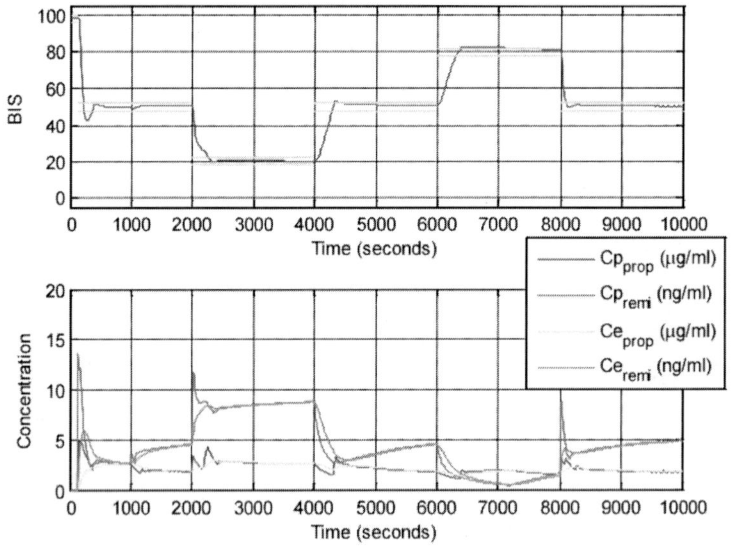

Figure 21. BIS and drug concentration for testing profile by two T1 fuzzy PID controllers.

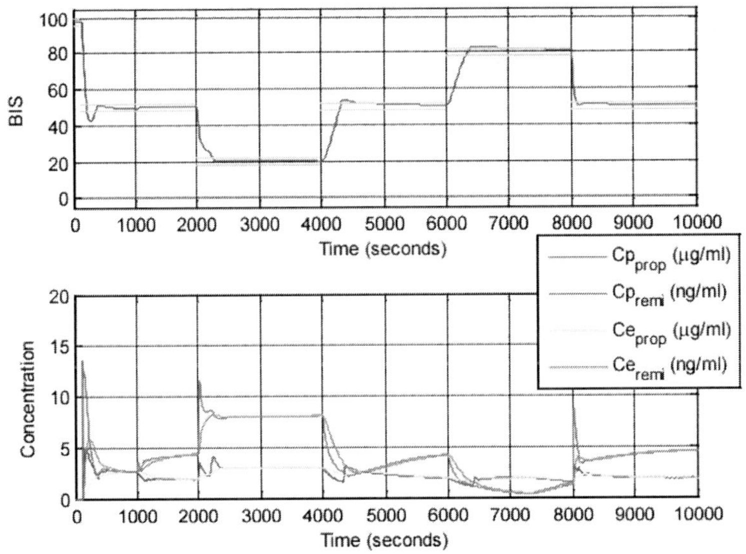

Figure 22. BIS and drug concentration for testing profile by two IT2 fuzzy PID controllers.

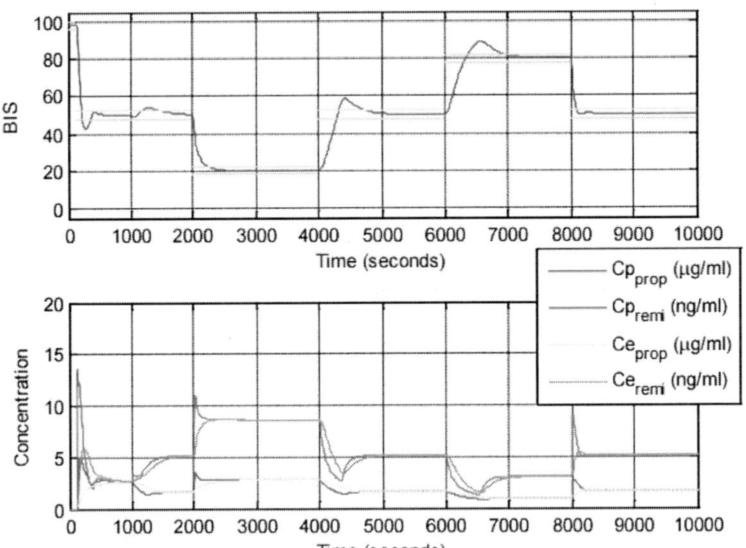

Figure 23. BIS and drug concentration for testing profile by one PID controller with scaling factors.

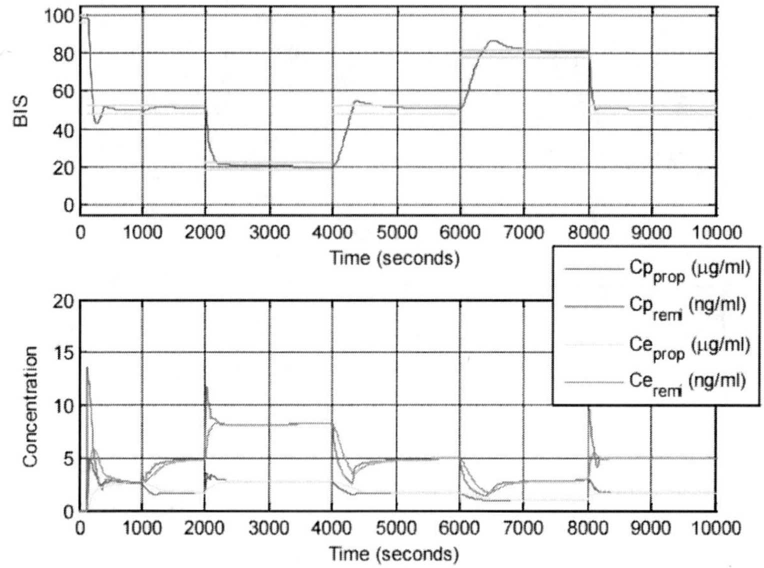

Figure 24. BIS and drug concentration for testing profile by one T1 fuzzy PID controller with T1 fuzzy scaling factors.

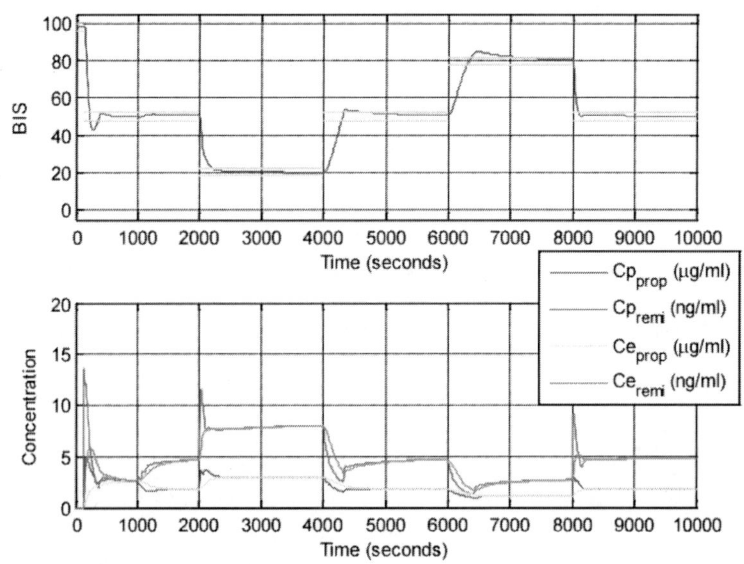

Figure 25. BIS and drug concentration for testing profile by one IT2 fuzzy PID controller with IT2 fuzzy scaling factors.

Table 15. The cost J for the testing profile.

Case	1	2	3
Cost	4.1175	3.9888	3.9745
Case	4	5	6
Cost	4.1228	4.0408	4.1066

Overall from a clinical point of view the results and performance obtained are acceptable, considering the lack of significant overshoot, the stability of the controller response and the ability of achieving the desired targets on a short time with clinically acceptable Ce. The controller also provided the expected response when the value of BIS target increases, and it stops infusion of anesthetic drugs but only restarts its infusion in other to avoid an overshoot.

When the PID control strategies with the best set of parameters (offering the best cost of J) are employed to regulate the BIS value using the testing profile in Figure 6, their cost values are listed in Table 15. The rank of which is identical to the sequence of cost values from the training profile. It indicates that the advantage of fuzzy PID controller over linear PID controller exists not only in the training profile, but also in various target profiles.

6. CONCLUSION

In this paper, drug administration for anesthesia has been realized by various PID control strategies. We have first constructed a multivariable anesthesia model with propofol and remifentanil based on 42 patients' clinical data. Simulation has been conducted to regulate the output BIS value governed by this model using six control strategies including linear PID controllers, T1 fuzzy PID controllers and IT2 fuzzy PID controllers. These six strategies are separated into two groups, namely two controllers and one controller with scaling factors, to handle the co-administration of two drugs. Parameters of these controllers have all been optimized by GA subject to a performance index quantitatively measuring the control performance. In order to make the control sophistica- ted, target profiles have been designed for BIS regulation. Different target profiles have been utilized to test and verify the performance of controllers. It has been demonstrated that the IT2 fuzzy PID controllers offer the best performance, and T1 fuzzy PID controllers come to the second. In the future, further investigation can be carried out to control a general and non-parametric multivariable anesthesia model rather than the model with parameters obtained from specific group of patients.

ACKNOWLEDGEMENTS

The work described in this paper was partly supported by King's College London, Fundação para a Ciênciae Tecnologia (grant number SFHR/BD/44162/2008) and European Social Fund in the POPH framework.

REFERENCES

1. Croft, R. and Washington, S. (2012) Induction of Anaesthesia. Anaesthesia & Intensive Care Medicine, 13, 401-406.
2. Mortier, E.P. and Struys, M.M. (2001) Monitoring the Depth of Anaesthesia Using Bispectral Analysis and Closed- Loop Controlled Administration of Propofol. Best Practice & Research Clinical Anaesthesiology, 15, 83-96.
3. Biebuyck, J.F., O'Hara, D.A., Bogen, D.K. and Noordergraaf, A. (1992) The Use of Computers for Controlling the De- livery of Anesthesia. Anesthesiology, 77, 563-581.
4. Kern, S.E., Xie, G., White, J.L. and Egan, T.D. (2004) A Response Surface Analysis of Propofol-Remifentanil Pharmacodynamic Interaction in Volunteers. Anesthesiology, 100, 1374-1381.
5. Struys, M., De Smet, M.T., Greenwald, S., Absalom, A.R., Bingé, S. and Mortier, E.P. (2004) Performance Evaluation of Two Published Closed-Loop Control Systems Using Bispectral Index Monitoring: A Simulation Study. Anesthesio- logy, 100, 640-647.
6. Struys, M.M., Mortier, E.P. and Smet, T.D. (2005) Closed Loops in Anaesthesia. Best Practice & Research Clinical A- naesthesiology, 20, 211-220.
7. Bibian, S., Ries, C.R., Huzmezan, M. and Dumont, G. (2005) Introduction to Automated Drug Delivery in Clinical A- nesthesia. European Journal of Control, 11, 535-557.

8. Moppett, I. (2008) Inhalational Anaesthetics. Anaesthesia & Intensive Care Medicine, 9, 567-572.

9. Kim, D.W., Joo, J.D., In, J.H., Jeon, Y.S., Jung, H.S., Jeon, K.B., Park, J.S. and Choi, J.W. (2013) Comparison of the Recovery and Respiratory Effects of Aminophylline and Doxapram Following Total Intravenous Anesthesia with Propofol and Remifentanil. Journal of Clinical Anesthesia, 25, 173-176.

10. Hawthorne, C. and Sutcliffe, N. (2013) Total Intravenous Anaesthesia. Anaesthesia & Intensive Care Medicine, 14, 129-131.

11. Westrin, P. (1996) 6 Intravenous and Inhalational Anaesthetic Agents. Baillière's Clinical Anaesthesiology, 10, 687- 715.

12. Grieder, P., Gentilini, A., Morari, M. and Schnider, T. (2001) Robust Adaptive Control of Hypnosis during Anesthesia. Proceedings of the 23rd Annual International Conference of the IEEE Engineering in Medicine and Biology Society, 2, 2055-2058.

13. Kissin, I. (1997) A Concept for Assessing Interactions of General Anesthetics. Anesthesia and Analgesia, 85, 204-210.

14. Liu, J., Singh, H. and White, P.F. (1997) Electroencephalographic Bispectral Index Correlates with Intraoperative Recall and Depth of Propofol-Induced Sedation. Anesthesia and Analgesia, 84, 185-189.

15. Glass, P.S., Bloom, M., Kearse, L., Rosow, C., Sebel, P. and Manberg, P. (1997) Bispectral Analysis Measures Sedation and Memory Effects of Propofol, Midazolam, Isoflurane and Alfentanil in Healthy Volunteers. Anesthesiology, 86, 836-847.

16. Panagopoulos, H., Astrom, K. and Hagglund, T. (2002) Design of PID Controllers Based on Constrained Optimization. IEEE Proceedings of Control Theory and Applications, 149, 32-40.

17. Dumont, G.A. Martinez, A. and Ansermino, J.M. (2009) Robust Control of Depth of Anesthesia. International Journal of Adaptive Control and Signal Processing, 23, 435-454.

18. Soltesz, K., Hahn, J.O., Hägglund, T., Dumont, G.A. and Ansermino, J.M. (2013) Individualized Closed-Loop Control of Propofol Anesthesia: A Preliminary Study. Biomedical Signal Processing and Control, 8, 500-508.

19. Shieh, J.S., Kao, M.H. and Liu, C.C. (2006) Genetic Fuzzy Modelling and Control of Bispectral Index (BIS) for General Intravenous Anaesthesia. Medical Engineering & Physics, 28, 134-148.

20. Sartori, V., Schumacher, P., Bouillon, T., Luginbuehl, M. and Morari, M. (2005) On-Line Estimation of Propofolphar- macodynamic Parameters. Proceedings of IEEE-EMBS 2005, 27th Annual International Conference of the Engineering in Medicine and Biology Society, 74-77.

21. De Smet, T., Struys, M.M., Neckebroek, M.M., Van den Hauwe, K., Bonte, S. and Mortier, E.P. (2008) The Accuracy and Clinical Feasibility of a New Bayesian-Based Closed-Loop Control System for Propofol Administration Using the Bispectral Index as a Controlled Variable. Anesthesia and Analgesia, 107, 1200-1210.

22. Hahn, J.O., Dumont, G.A. and Ansermino, J.M. (2012) Robust Closed-Loop Control of Hypnosis with Propofol Using WAV_{CNS} Index as the Controlled Variable. Biomedical Signal Processing and Control, 7, 517-524.

23. Haddad, W.M., Hayakawa, T. and Bailey, J.M. (2006) Adaptive Control for Nonlinear Compartmental Dynamical Sys- tems with Applications to Clinical Pharmacology. Systems & Control Letters, 55, 62-70.

24. Agrawal, D., Kumar, S., Kumar, A., Gombar, S., Trikha, A. and Anand, S. (2012) Design of an Assistive Anaesthesia Drug Delivery Control Using Knowledge Based Systems. Knowledge-Based Systems, 31, 1-7.

25. Mahfouf, M., Nunes, C.S., Linkens, D.A. and Peacock, J.E. (2005) Modelling and Multivariable Control in Anaesthesia Using Neural-Fuzzy Paradigms: Part II. Closed-Loop Control of Simultaneous Administration of Propofol and Remifentanil. Artificial Intelligence in Medicine, 35, 207-213.

26. El-Bardini, M. and El-Nagar, A.M. (2011) Direct Adaptive Interval Type-2 Fuzzy Logic Controller for the Multivariable Anaesthesia System. Ain Shams Engineering Journal, 2, 149-160.

27. Marsh, B., White, M., Morton, N. and Kenny, G.N. (1991) Pharmacokinetic Model Driven Infusion of Propofol in Chi- ldren. British Journal of Anaesthesia, 67, 41-48.

28. Minto, C.F., Schnider, T.W., Egan, T.D., Youngs, E., Lemmens, H.J., Gambus, P.L., Billard, V., Hoke, J.F., Moore, K.H., Hermann, D.J., Muir, K.T., Mandema, J.W. and Shafer, S.L. (1997) Influence of Age and Gender on the Pharmacokinetics and Pharmacodynamics of Remifentanil. I. Model Development. Anesthesiology, 86, 10-23.

29. Absalom, A. and Struys, M. (2007) An Overview of TCI & TIVA. Academia Press, Gent.

30. Minto, C.F., Schnider, T.W., Short, T.G., Gregg, K.M., Gentilini, A. and Shafer, S.L. (2000) Response Surface Model for Anesthetic Drug Interactions. Anesthesiology, 92, 1603-1616.

31. Bruhn, J., Bouillon, T.W., Radulescu, L., Hoeft, A., Bertaccini, E. and Shafer, S.L. (2003) Correlation of Approximate Entropy, Bispectral Index, and Spectral Edge Frequency 95 (SEF95) with Clinical Signs of "Anesthetic Depth" during Coadministration of Propofol and Remifentanil. Anesthesiology, 98, 621-627.

32. Alvis, J.M., Reves, J.G., Spain, J.A. and Sheppard, L.C. (1985) Computer-Assisted Continuous Infusion of the Intravenous Analgesic Fentanyl during Generalanesthesia—An Interactive System. IEEE Transactions Biomedical Engineering, 32, 323-329.

33. Ang, K.H., Chong, G. and Li, Y. (2005) PID Control System Analysis, Design, and Technology. IEEE Transactions on Control Systems Technology, 13, 559-576.

34. Carvajal, J., Chen, G. and Ogmen, H. (2000) Fuzzy PID Controller: Design, Performance Evaluation, and Stability A- nalysis. Information Sciences, 123, 249-270.

35. Çetin, Ş. and Akkaya, A.V. (2010) Simulation and Hybrid Fuzzy-PID Control for Positioning of a Hydraulic System. Nonlinear Dynamics, 61, 465-476.

36. Hagras, H. (2007) Type-2 FLCs: A New Generation of Fuzzy Controllers. IEEE Computational Intelligence Magazine, 2, 30-43.

37. John, R. and Coupland, S. (2007) Type-2 Fuzzy Logic: A Historical View. IEEE Computational Intelligence Magazine, 2, 57-62.

38. Lam, H.K. and Seneviratne, L.D. (2008) Stability Analysis of Interval Type-2 Fuzzy-Model-Based Control Systems. IEEE Transactions on Systems, Man, and Cybernetics, Part B: Cybernetics, 38, 617-628.

39. Lam, H.K., Li, H., Deters, C., Wuerdemann, H., Secco, E. and Althoefer, K. (2014) Control Design for Interval Type-2 Fuzzy Systems under Imperfect Premise Matching. IEEE Transactions on Industrial Electronics, 61, 956-968.

40. Mendel, J.M., John, R.I. and Liu, F. (2006) Interval Type-2 Fuzzy Logic Systems Made Simple. IEEE Transactions on Fuzzy Systems, 14, 808-821.

41. Mendel, J.M. (2007) Type-2 Fuzzy Sets and Systems: An Overview. IEEE Computational Intelligence Magazine, 2, 20-29.

42. Du, X. and Ying, H. (2010) Derivation and Analysis of the Analytical Structures of the Interval Type-2 Fuzzy-PI and PD Controllers. IEEE Transactions on Fuzzy Systems, 18, 802-814.

3

CHAPTER

OPTIMAL ENERGY MANAGEMENT, LOCATION AND SIZE FOR STATIONARY ENERGY STORAGE SYSTEM IN A METRO LINE BASED ON GENETIC ALGORITHM

Huan Xia *, Huaixin Chen, Zhongping Yang, Fei Lin and Bin Wang

School of Electrical Engineering, Beijing Jiaotong University, No.3 Shangyuancun, Beijing 100044, China

ABSTRACT

The installation of stationary super-capacitor energy storage system (ESS) in metro systems can recycle the vehicle braking energy and improve the pantograph voltage profile. This paper aims to optimize the energy management, location, and size of stationary super-capacitor ESSes simultaneously and obtain the best economic efficiency and voltage profile of metro systems. Firstly, the simulation platform of an urban rail power supply system, which includes trains and super-capacitor energy storage systems, is established. Then, two evaluation functions from the perspectives of economic efficiency and voltage drop compensation are put forward. Ultimately, a novel optimization method that combines genetic algorithms and a simulation platform of urban rail power supply system is proposed, which can obtain the best energy management strategy, location, and size for ESSes simultaneously. With actual parameters of a Chinese metro line applied in the simulation comparison, certain optimal scheme of ESSes' energy management strategy, location, and size obtained by a novel optimization method can achieve much better performance of metro systems from the perspectives of two evaluation functions. The simulation result shows that with the increase of weight coefficient, the optimal energy management strategy, locations and size of ESSes appear certain regularities, and the best compromise between economic efficiency and voltage drop compensation can be obtained by a novel optimization method, which can provide a valuable reference to subway company.

Keywords: energy storage system; super-capacitor; energy management; configuration; economic efficiency; voltage drop compensation; genetic algorithm

1. INTRODUCTION

In recent years, with the rapid development of the Chinese economy, growing environmental pollution, and traffic congestion in major cities are becoming serious social issues. For the purpose of improving the urban environment and energy efficiency, the development of modern urban rail transit, which has the significant advantages of large capacity, punctuality, safety, energy conservation, and environmental protection, becomes a social consensus [1,2]. Low running resistance and the reuse of braking energy are two main factors that make urban rail transit better than other means of transport in energy efficiency. Recent studies have shown that up to 40% of the energy supplied to electrical rail guided vehicles could be recovered through regenerative braking [1]. In a metro network system, the trains are accelerated and braked frequently. Since most of the rectifiers in the metro network are unidirectional, the regenerative braking energy cannot be returned to the supply network, and if there are no adjacent accelerating trains or energy storage system to absorb the regenerative energy, the surplus braking energy has to be wasted on the mechanical braking or on-board resistors. If different trains are close to each other and they start all together, contact lines will become overloaded and the pantograph voltages of trains will drop significantly, which results in high lines loss and the opening of minimum voltage protective action of trains by limiting the current. Hence, the installation of energy storage systems in urban railway transit has become a universal concern, which can recycle the regenerative braking energy, prevent regeneration cancellation, shave the peak power of substations, and compensate the voltage drops of pantograph quickly.

Current research activities have presented the application of batteries, flywheels, super-capacitors, and hybrid energy storages as energy storage devices [3,4,5,6,7,8]. Among the different storage systems available, super-capacitors seem to be the most appropriate for the application in a metro system for the advantages of rapid charging and discharging frequencies, a long cycle life, and high power density, which highly match the characteristics of metro system, such as short running time between stations, frequent accelerating and braking, booming power within a short time, *etc.* Super-capacitor energy storage systems (ESS) can be either stationary or on-board [8,9,10,11]. The allocation on board of the storage system increases the train mass and requires additional space for their accommodation. Thus, stationary ESSes set inside traction substations (TSSs) are preferred for metro systems, and their best energy management, location, and size will be discussed in this paper.

Several papers have dealt in depth with optimization of energy management strategies of stationary ESSes [12,13,14]. Among them [12] proposes a control strategy based on the maximum kinetic energy recovery throughout braking operations of the running vehicles. The strategy stays on the knowledge of the state of charge of ESS and the actual vehicle speeds. Reference [13] proposes a optimization procedure based on a linearized modeling of the electrical LRV network, the target of the control strategy is the optimal tracking of the storage device voltage subject to the minimization of the substations supplied power. Optimal location and size of ESSes are also investigated in detail in [15,16,17,18,19,20,21,22]. Reference [15] discusses the configuration of ESSes for voltage drop compensation, which takes account of the topology of the line and the movement of the vehicles. Reference [16] proposes an optimization method based on a genetic algorithm, which can obtain certain preferable location and size for ESSes.

However, there are still some drawbacks on the above research. Firstly, some of the references involve only small amounts of substations and vehicles when modeling the urban railway network [12,13,14,15] and some of them do not take into account the time-variation (network topology change with vehicle movement) and nonlinearity (nonlinearity of substation and regenerative braking) of the network structure. Secondly, and most importantly, the optimization research of energy management strategy and configuration for ESSes will influence each other, and they both affect the performances of urban railway network, while the configuration optimization research of ESSes in reference

[16,17,18,19,20,21,22] is on the premise that energy management strategy of ESSes is fixed and invariable.

In this paper improved energy management strategy of ESSes and novel optimization method are proposed. Compared to previous work [16,17,18,19,20,21,22], the improved energy management strategy can manage and coordinate the energy flow of multiple ESSes, which can achieve smoother changes of voltages and currents in the system and improve the energy savings of ESSes effectively, and the new proposed optimization algorithm can further improve the performance of ESSes by optimizing energy management parameters, location, and size of ESSes simultaneously, which has rarely been studied in previous work about the optimization of ESSes, and the evaluation functions of proposed optimization algorithm in this paper are more appropriate, which are put forward from the perspectives of economic efficiency and voltage drop compensation.

The organization of this work is as follow: the simulation platform of urban rail power supply system, which includes trains and super-capacitor energy storage systems has been established in Section 2; additionally, particular data of the researched Chinese metro line is given. Then Section 3 sets up two evaluation functions from the perspectives of economic efficiency and voltage drop compensation. InSection 4, a novel optimization method based on genetic algorithm (GA) is put forward, which can optimize energy management strategy, location and size of ESSes simultaneously. Finally in Section 5, the result of the simulation comparison is presented and discussed.

2. MODELING

2.1. Model of Metro Power Supply Network

The model of metro system's DC traction power supply network is shown in Figure 1 [16]. In order to show the behavior of the metro power supply network as correctly as possible, all components of the metro network, which includes irreversible traction substations (TSS), trains, metro lines, and stationary energy storage systems (ESSes), will be modeled appropriately to maintain original characters of the network structure's time-variation and nonlinearity.

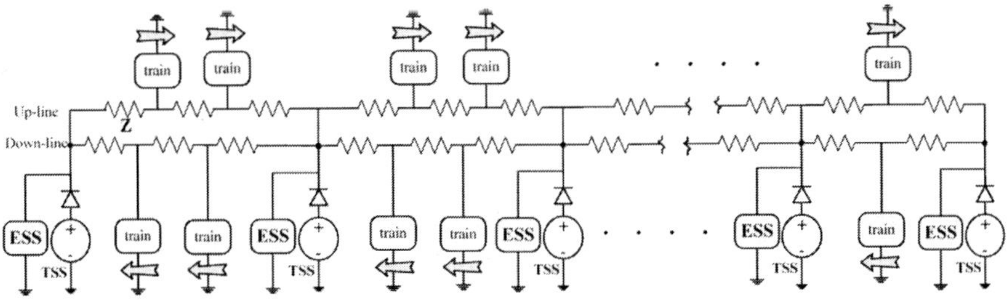

Figure 1. The model of metro system's DC traction power supply network.

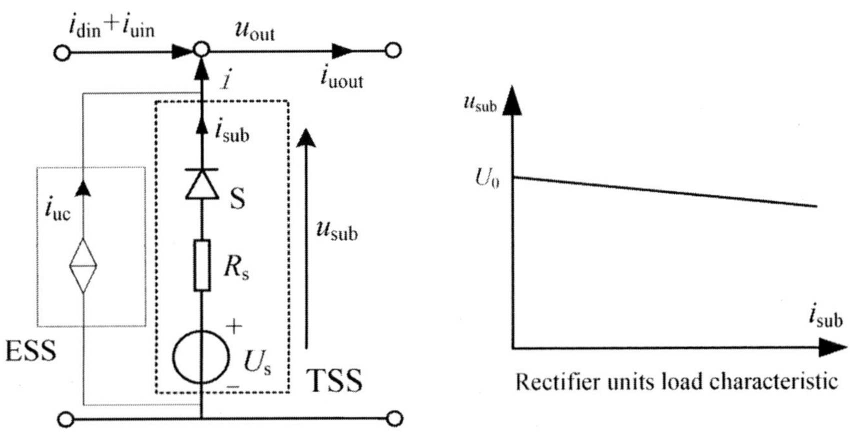

Figure 2. TSS model.

2.1.1. Traction Substation (TSS) Model

As shown in Figure 2, the substation is modeled by an ideal DC voltage source connected in series with its equivalent internal resistance R_S and the diode D, which to simulate output characteristics. When the output current of substation is increased, the voltage of substation decreases correspondingly to limit its output power. U_0 is the no-load voltage of substations.

$$i_{uout} = i_{uin} + i_{din} + i \tag{1}$$

$$i = i_{uc} + i_{sub} \tag{2}$$

$$i_{\text{sub}} \geq 0 \tag{3}$$

2.1.2. Train Model

As shown in Figure 3, the train model is modeled by a controlled current source which draws electric power at the accelerating time and delivers braking power at the regenerative time. The impedance of the line connected to the trains is expressed as *Z*, the value of which is time-varying, and it is linear with the line length that is determined by the present position of trains. When pantograph voltage exceeds U_b, the braking resistor R_b will consume the braking energy. R_f is vehicle filter resistance; L_f is vehicle filter inductance; C_{fc} is the support capacitor of train; P_{aux} is auxiliary power; and P is the electric power of train.

$$u_{\text{out}} = u_{\text{in}} + Ri_{\text{out}} + L\frac{di_{\text{out}}}{dt} \tag{4}$$

$$i_{\text{out}} = i_{\text{in}} + i \tag{5}$$

$$i = -i_{inv} - \frac{p_{aux}}{u_{fc}} - C_{fc}\frac{du_{fc}}{dt} \tag{6}$$

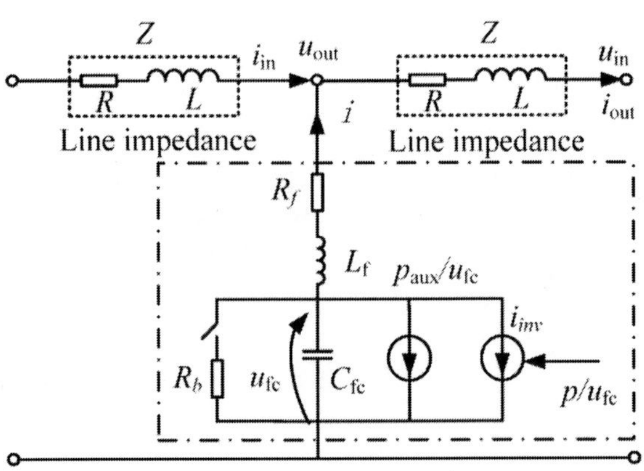

Figure 3. Train model.

2.1.3. Energy Storage System (ESS) Model

The ESS model consists of the super-capacitors, controlled current source, and energy management strategy controller, is shown in Figure 4. The ESS model is connected in parallel with the output of the substation, and it can deliver or draw the electric power from the metro power supply network through the current source which is controlled by the energy management strategy and configuration of super-capacitors in real time.

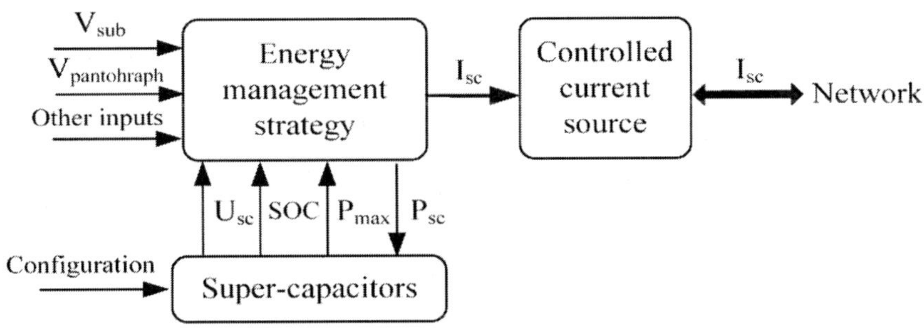

Figure 4. Stationary ESS model.

The SOC (State of Charge) of super-capacitors is defined as follows, it represents the storage energy of ESS, which is proportional to the square of the terminal voltage.

$$0.25 \le SOC = \frac{E_{sc}}{E_{scmax}} = \frac{0.5CU_{sc}^2}{0.5CU_{scmax}^2} = \frac{U_{sc}^2}{U_{scmax}^2} \le 1 \tag{7}$$

In a practical application, the function of controlled current source in the model is generally implemented by the unidirectional DC/DC converter. In order to maintain the normal operation of DC/DC converter, the terminal voltage of super-capacitors should be set between 50% and 100% the maximum voltage, so the range of SOC varies from 0.25 to 1.

2.2. Simulation Platform of Metro System for Power Flow Calculation

As above, the model of a DC metro power supply network (DC-PSN) is set up by a novel approach of component segmentation. In order to calculate the power flow of the DC metro power supply network, an integrated simulation platform, which includes DC metro power supply network (DC-PSN), train performance simulator (TPS), and super-capacitor energy storage system (SCESS) is established in the Matlab environment, as shown in Figure 5 [16].

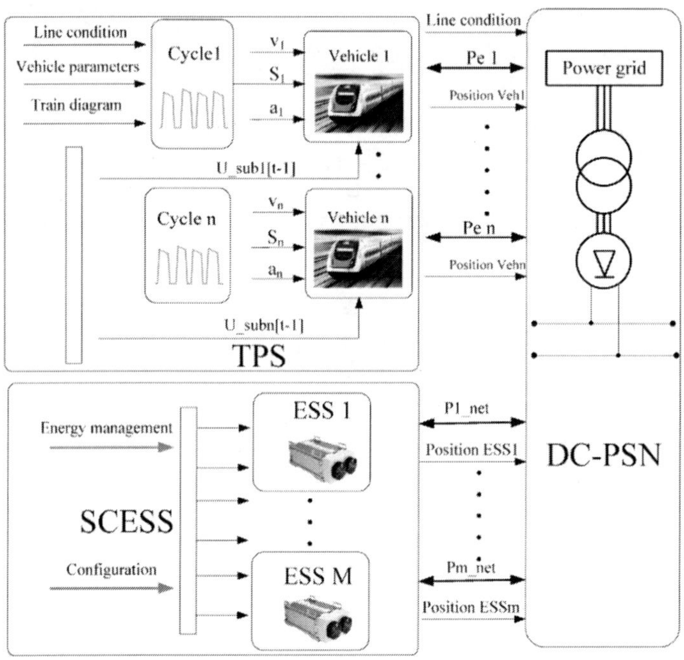

Figure 5. Simulation platform of metro system for power flow calculation.

2.2.1. DC Metro Power Supply Network (DC-PSN)

In the previous section, the paper has presented the structure and model of a DC-PSN. In the power flow calculation of the DC metro power supply network, because of its time-variation (network topology changes with train movement) and nonlinearity (nonlinearity of substation and regenerative braking) of the network structure, a new power flow calculation method by component segmentation is presented. The simulation result shows excellent rapidity and astringency can be obtained by this method. Moreover, the structure and model of the DC-PSN can be extended easily.

2.2.2. Train Performance Simulator (TPS)

As shown in Figure 5, the output of TPS is not only associated with line condition, vehicle data, and timetable, but is also constrained by real-time train pantograph voltage. From the TPS we can get positions of up-line and down-line trains and their corresponding electric power, which offer essential data for subsequent power flow calculation of the DC supply network.

2.2.3. Super-Capacitor Energy Storage System (SCESS)

SCESS set certain energy management strategy, location, and size of ESSes on different substations, which determine the power direction and value of ESSes in real time. The installation of ESSes will change the power flow of the DC metro power supply network and the system performances can be improved significantly by setting the most appropriate energy management strategy, location, and size.

2.3. Case Data

A particular case of Beijing Subway line is studied in this paper. The total length of the line is about 11.3 km along with 12 stations, of which seven are traction substations and their distribution is shown in Table 1. The vehicle data and metro DC network parameters are shown as Table 2. These parameters are provided by the Beijing Subway Company.

Table 1. TSS spacing distances.

Traction substation	1–2	2–3	3–4	4–5	5–6	6–7
Substation spacing (km)	1.1	1.9	2.2	2.3	2.1	2.7

Table 2. Vehicle data.

Parameter	Value	Parameter	Value
Formation	3M3T	Inverter efficiency	0.97
Load condition	312.9t (AW3)	Motor efficiency	0.915
Rated voltage	750 Volt	Gearing efficiency	0.93
AC motor/M	180 kW × 4	Max speed	80 km/h
SIV power	160 kVA × 2	Max acceleration	1 m/s^2
SIV power factor	0.85	Min deceleration	−1 m/s^2
Floating voltage U_s	836 Volt	Equivalent internal resistance R_s	0.07 Ω
Contract line impendence	0.007 Ω/km	Rail impendence	0.009 Ω/km
Pantograph impendence	0.015 Ω	–	–

2.4. Simulation Output

Under simulation conditions, super-capacitor ESSes of 14 kWh are configured in every other substation and controlled with a traditional double-loops control strategy [16]. The simulation output waveforms is shown in Figure 6, which include speed and electric power of an up-line train, voltage and current of the train pantograph, voltage and current of a substation, charging energy, and SOC of the ESS in the substation. SOC of ESS varies between 0.25 and 1.

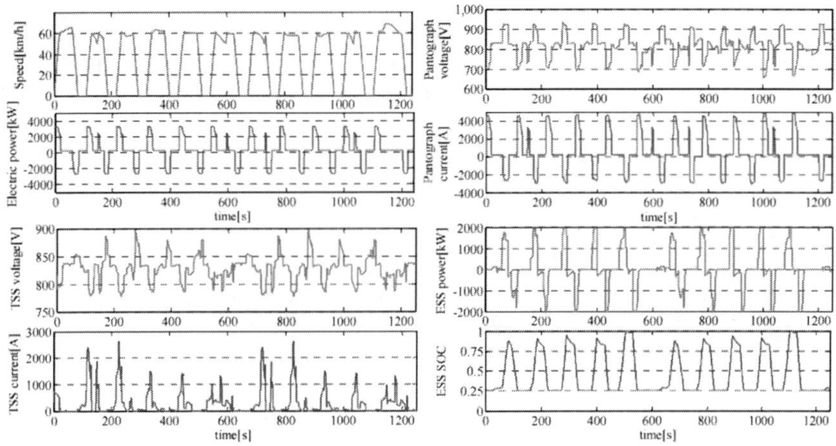

Figure 6. The output waveforms of simulation platform.

3. OBJECTIVE FUNCTION

3.1. Objective Function

In order to evaluate the system performances in terms of energy saving, voltage drop compensation, and installation cost for different energy management strategy and configuration of ESSes, the paper puts forward two evaluation functions and one objective function.

3.1.1. Economic Efficiency, e%

Economic efficiency e% is put forward from the viewpoint of considering energy savings and installation cost in a unified way to evaluate the economic return rates of ESSes for Subway Company. Economic efficiency e% is a percentage calculated by dividing the total electricity price of the substations by economic savings (returns minus costs).

As shown in Figure 7, one super-capacitor ESS, which includes the connection unit, DC/DC converter, and super-capacitor strings, is installed on the traction substation. The circuit structure of super-capacitor ESS is shown in Figure 8. The installation cost of ESSes is determined by various factors, which include the capacity, equipment, control circuit, maintenance cost, *etc.* The cost of DC/DC converters and super-capacitor strings are determined mainly by the maximum power of the ESS. In order to ease the DC/DC converter design, maximum voltage of super-capacitor strings should be lower than network voltage (836 V at no load). Hence, six super-capacitor modules (BMOD0063P125) are put in series to form a super-capacitor string, which has terminal voltage of 750 V and maximum continuous power of 180 kW, and the configuration of super-capacitors installed in every substation could be adjusted by changing the number of paralleled super-capacitor strings. The Parameters of super-capacitor modules (BMOD0063P125) are shown as Table 3.

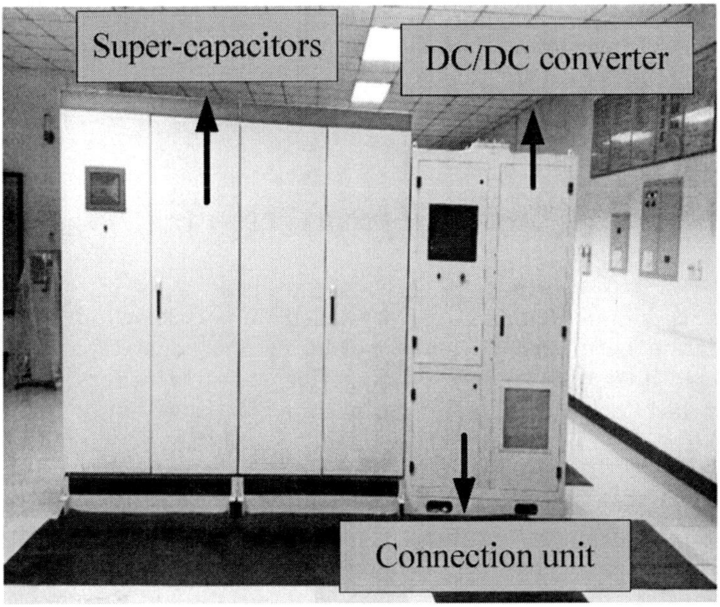

Figure 7. Super-capacitor ESS installed in traction substation.

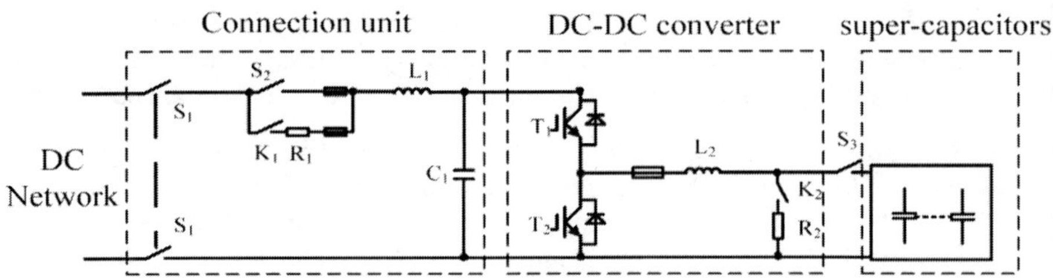

Figure 8. The circuit structure of super-capacitor ESS.

Table 3. Parameters of super-capacitor module (BMOD0063P125).

Parameter	Value	Parameter	Value
Rated voltage	125 V	Capacitance	63 F
Maximum continuous current	240 A	Maximum continuous power	30 kW
Maximum ESR_{DC}, initial	0.018 Ω	Energy	0.137 kWh
Price	5,333 $		

The cost of investment for a super-capacitor ESS on substation *k* during their life time of *l* years, can be calculated by:

$$Cost_k = \begin{cases} 0, n_k = 0 \\ (C + n_k \times p \times m) \times (1 + r)^l, 0 < n_k \leq 18 \end{cases} \tag{8}$$

where n_k is the number of paralleled super-capacitor strings on substation *k*; *p* is the maximum power of one super-capacitor string; and *m* is dollar per power constant for super-capacitors and DC/DC converter. If n_k equals 0, a super-capacitor ESS would not be installed on substation *k*. If n_k is more than 0, the cost of investment for super-capacitor ESS $Cost_k$ includes two parts. nk×p×m is the cost of DC/DC converters and super-capacitor strings that are determined by the maximum power of the ESS; *C* is other part of installation cost from protective device, breaker, maintenance cost, *etc.*, which has a small relationship to power of the ESS. *r* is the rate of return constant. Considering the limited free space of each metro substation, the number of paralleled super-capacitor strings n_k on each substation is no bigger than 18 in this paper.

By taking the sum of output energy consumption of all TSSes along the metro line, the total energy consumption of the substations in kWh during one year can be calculated from the following formula:

$$E_{\text{sub}} = \sum_{1}^{k} \left[\int_{0}^{T} (I_{\text{sub}} \cdot U_{\text{sub}}) dt \right] \times \frac{365}{3600000} \quad (9)$$

where k is the number of traction substations; T is the running time in one day. U_{sub}, I_{sub} are, respectively, the voltage and current of substation.

The application of ESSes in a metro system can reduce the total energy consumption of the substations because of the recycle of trains' regenerative braking energy, but the installation cost should also be considered as well. The total profit obtained by ESSes in l years should be the difference between the saved electricity price and the installation cost of ESSes.

$$P_{\text{sub}}^{\text{nosc}} = E_{\text{sub}}^{\text{nosc}} \times \varepsilon \times \frac{l(2+(l-2)i)}{2} \quad (10)$$

where Enoscsub is one year energy consumption of the substations in absence of ESSes; ε is electricity price in dollar per kWh, Pnoscsub is l years' electricity price of the substations in absence of ESSes, and i is the yearly inflation of electricity price.

$$P_{\text{sub}}^{\text{sc}} = E_{\text{sub}}^{\text{sc}} \times \varepsilon \times \frac{l(2+(l-2)i)}{2} + \sum_{1}^{k} Cost_k \quad (11)$$

where Escsub is one year' energy consumption of the substations in presence of ESSes, Costk is installation cost of ESS on substation k, and Pscsub is l years' expenditure of the substations in presence of ESSes, which includes the electricity price and installation cost of ESSes.

In this paper, economic efficiency e% is defined as the following formula:

$$e\% = \frac{P_{\text{sub}}^{\text{nosc}} - P_{\text{sub}}^{\text{sc}}}{P_{\text{sub}}^{\text{nosc}}} \times 100\% \quad (12)$$

when economic efficiency e% equals 0, it means the saved electricity price is the same as the installation cost of ESSes. Necessary parameters for calculating the economic efficiency e% of ESSes are given in Table 4.

Table 4. Necessary parameters for calculating economic efficiency.

Parameter	Value	Parameter	Value
p	180 kW	ε	0.16 $/kWh
m	0.244 $/W	r	5%
C	0.16 M$	i	5%
l	10 years		

3.1.2. Voltage Drop Compensation, v%

If different trains are close to each other and they start all together, contact lines will become overloaded and the pantograph voltages of trains will drop significantly, which results in high line loss and the opening of minimum voltage protective action of trains by limiting the current. The installation of ESSes in the metro system can shave the peak power of substations, improve the load capacity of the system, and compensate the pantograph drops quickly. Voltage drop compensation v%, in this paper, evaluates in percent the voltage drop compensation at the pantograph, giving the rate about how much the voltage drops improvement is when the ESSes are installed.

$$v\% = \frac{\sum_1^j \int_{U_p<U_r} \left(U_r - U_p^{nosc}\right)dt - \sum_1^j \int_{U_p<U_r} \left(U_r - U_p^{sc}\right)dt}{\sum_1^j \int_{U_p<U_r} \left(U_r - U_p^{nosc}\right)dt} \times 100\% \tag{13}$$

where, *Up* is the pantograph voltage of trains; *Ur* is the rated voltage of trains' pantograph; *j* is the amount of up-line and down-line trains. From the Equation (14), voltage drop compensation v% is calculated based on the integral of voltage drops improvement when *Up* is less than *Ur*, which is more appropriate and comprehensive than the maximum voltage drop compensation just in a moment in [15].

3.1.3. Objective Function, $ObjV$

Given economic efficiency e% and voltage drop compensation v%, the objective function for optimal energy management strategy and configuration of ESSes is shown as below:

$$ObjV = \omega \times e\% + (1-\omega) \times v\% \quad (14)$$

where ω is the weight coefficient of economic efficiency e%, it represents the emphasis degrees of economic efficiency e%. When ω is set to 1, it means that economic efficiency is the only evaluation index considered in the optimization.

4. NOVEL OPTIMIZATION METHOD BASED ON A GENETIC ALGORITHM

The traditional optimization method based on a genetic algorithm proposed in [16] can optimize the location and size of ESSes significantly, but the adopted energy management strategy is constant. The energy management is also important for the performance improvement of a metro supply network, and the optimization of energy management and configuration for ESSes will influence each other. Thus, the proposed novel optimization method in this paper, which combines a genetic algorithm and a simulation platform of urban rail power supply system, is meant to optimize energy management, location, and size of ESSes simultaneously.

4.1. Improved Energy Management Strategy

In order to improve economic efficiency e% and voltage drop compensation v% of ESSes, an improved energy management strategy is put forward, which decides the charging and discharging current of the ESS by detecting the voltage of substation, ESS and train pantographs, as shown in Figure 9.

The improved energy management strategy can be divided into three parts: SOC constraint, current loop, and energy management. Due to the function of the SOC constraint, the working range of the SOC is 0.25–1, and the terminal voltage of ESSes is limited between 375 V and 750 V. Energy management can switch four work states to produce appropriate reference P_{sc}* for the ESS according to the substation voltage and pantograph voltage of the trains. The current loop can control the charging and discharging current of the super-capacitor ESS according to the reference I_{sc}*.

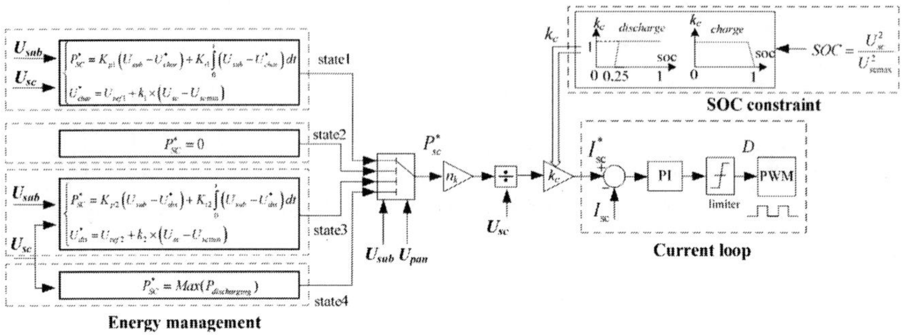

Figure 9. The improved energy management strategy of stationary ESSes.

The charging and discharging current reference I_{sc}* of the super-capacitor ESS can be calculated by Equation (15). And work states of super-capacitor ESS are shown in Figure 10.

$$I_{sc}^{*} = \frac{n_k \times k_c \times P_{sc}^{*}}{U_{sc}} \tag{15}$$

State 1: When the voltage of substation is higher than charging threshold value U_{char}*, the magnitude of the charging current reference P_{sc}* is determined by the PI controller according to the difference value between the present substation voltage and the threshold value U_{char}*. From Equation (16), if the electric braking power of train is small, the super-capacitor ESS will absorb all the regenerative braking energy and maintain the substation voltage at U_{char}*. Then, if the electric braking power of train is excessive, the super-capacitor ESS will absorb the braking energy with maximum charging current. The value of U_{char}* will increase with the increase of ESSes' terminal voltage, which enlarges the charging current of ESSes with smaller terminal voltage of ESSes significantly, as shown in Figure 11. The value of U_{scmin} is set to 375 V in this paper.

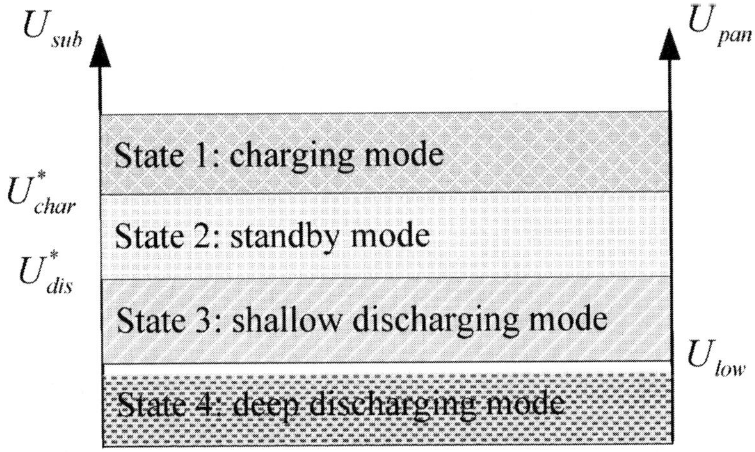

Figure 10. Work states of super-capacitor ESS.

$$\begin{cases} P_{sc}^* = K_{p1} \times \left(U_{sub} - U_{char}^*\right) + K_{i1} \times \int_0^t \left(U_{sub} - U_{char}^*\right) dt \\ U_{char}^* = U_{ref1} + k_1 \times \left(U_{sc} - U_{sc\min}\right) \\ K_{p1} \geq 0, K_{i1} \geq 0, U_{char}^* > U_{no\_load}, k_1 \geq 0, U_{sc\max} \geq U_{sc} \geq U_{sc\min} \end{cases} \tag{16}$$

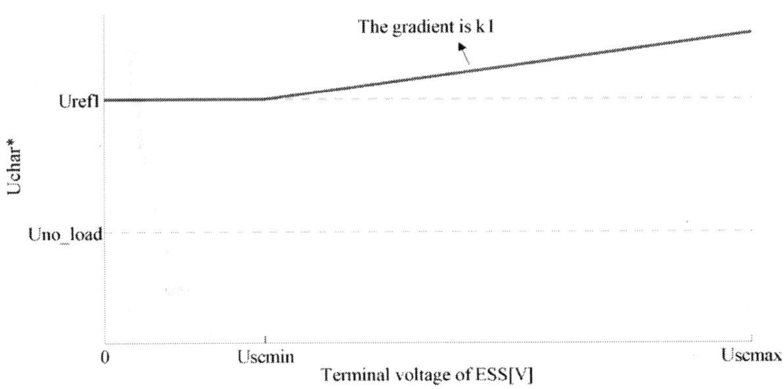

Figure 11. The values of U_{char}^*.

When the train is braking in one substation, by traditional energy management, the ESS installed in the substation will draw high power of regenerative energy and take no account of its terminal voltage and stored energy [16]. When the ESS is charged up to 100% with the regenerative energy, its terminal voltage would be 750 V and its charging current will be interrupted instantaneously, which leads to the drastic changes of substation current and line current, then all the regenerative energy of trains flows to the ESSes in the near substations as shown in Figure 12. ESSes are charged in turn and both with large current. On the contrary, by the improved energy management strategy, the ESS can adjust the threshold value U_{char}*according to its terminal voltage and achieve smoother changes of terminal voltage and charging current. Consequently, the regenerative energy is distributed to ESSes more evenly. It is worth mentioning that the improved energy management strategy reduces the line loss greatly and it also contributes to balance the terminal voltage for all different strings of ESS on a substation.

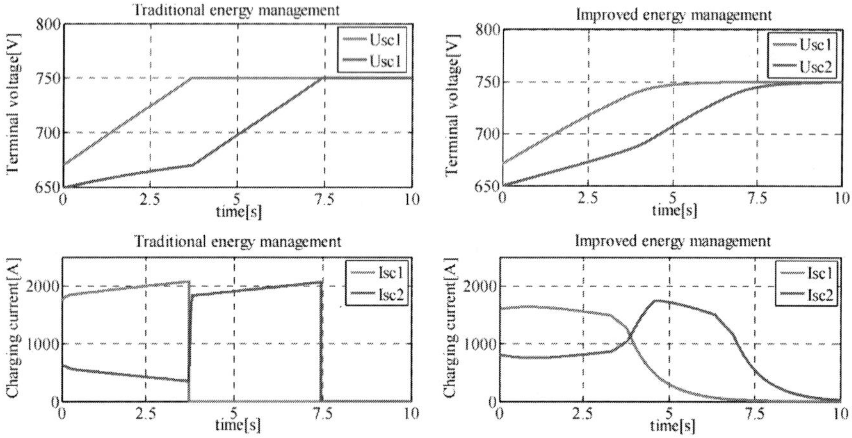

Figure 12. Terminal voltage and charging current of ESSes.

State 2: When the voltage of substation fluctuates between the charging threshold value U_{char}* and the discharging threshold value U_{dis}*, super-capacitor ESS maintain the standby state.

$$P_{SC}^{*} = 0 \tag{17}$$

State 3: When the voltage of substation is less than the discharging threshold value U_{dis}* and pantograph voltage of trains within one substation spacing range of ESS is higher than the low voltage threshold U_{low}, discharging power reference P_{sc}* of ESS is determined by the substation voltage U_{sub} and ESS terminal voltage U_{sc} simultaneously as follow:

$$\begin{cases} P_{SC}^{*} = K_{p2} \times \left(U_{sub} - U_{dis}^{*}\right) + K_{i2} \times \int_{0}^{t} \left(U_{sub} - U_{dis}^{*}\right) dt \\ U_{dis}^{*} = U_{ref2} + k_2 \times \left(U_{sc} - U_{sc\min}\right) \\ K_{p2} \geq 0, K_{i2} \geq 0, U_{dis}^{*} < U_{no\_load}, k_2 \geq 0, U_{sc\max} \geq U_{sc} \geq U_{sc\min} \end{cases} \tag{18}$$

The value of U_{dis}* will increase with the increase of ESSes' terminal voltage as shown in Figure 13, which enlarges the discharging current of ESSes with larger terminal voltage and balances SOC of ESSes significantly. When the accelerated train draws the energy in one substation, all ESSes nearby can deliver energy to shave the power of substations and compensate the voltage drops of the pantograph. As shown inFigure 14, by traditional energy management, the ESS installed in the substation will deliver highest power of energy and take no account of its terminal voltage and stored energy. When terminal voltage of one ESS decreases to U_{scmin}, its discharging current will be interrupted instantaneously, which also leads to drastic changes of substation current and line current. On the contrary, the improved energy management strategy can achieve smoother changes of voltages and currents in the system, and ESSes with higher SOC tend to deliver more energy to the supply network. Thus, by the improved energy management strategy, the flow of energy can be managed more steadily and effectively, and the line loss can be reduced greatly.

State 4: When the voltage of substation is less than dis charging threshold value U_{dis}* and the pantograph voltages of trains within one substation spacing range of ESS are less than the low voltage threshold U_{low}, super-capacitor ESS will deliver the energy with maximum discharging power. According to appropriate setting of K_{p2}, K_{i2}, k_2, U_{ref2}, U_{low}, ESS will retain proper energy when the pantograph voltage of a nearby train is acceptable, and when the pantograph voltage of a nearby train is very low, ESS delivers maximum discharging power to shave the peak power of the substation and compensate the pantograph voltage drop.

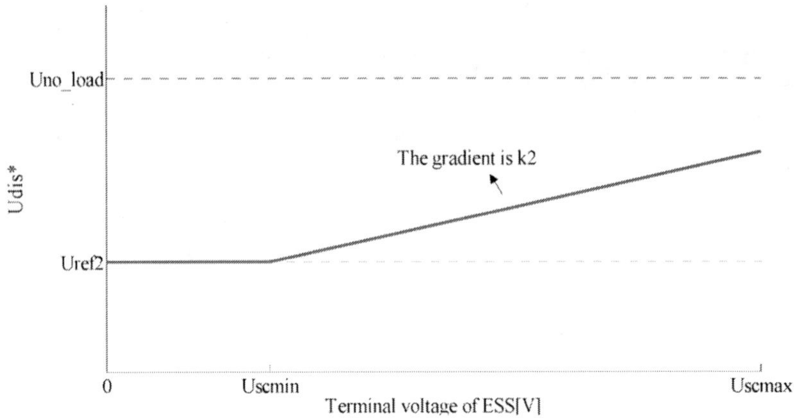

Figure 13. The values of U_{dis}^{*}.

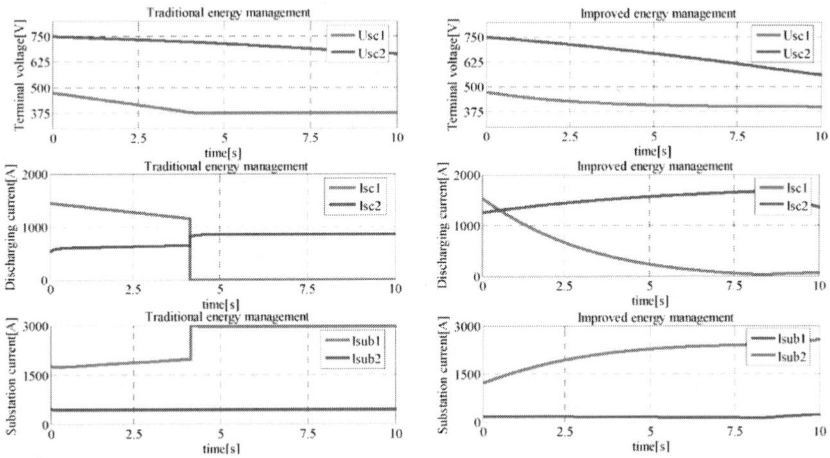

Figure 14. Terminal voltage and discharging current of ESSes.

In improved energy management strategy, K_{p1}, K_{i1}, k_1, K_{p2}, K_{i2}, k_2, U_{low}, U_{ref1}, U_{ref2} are nine undetermined parameters. In order to obtain best performance of system based on economic efficiency e%and voltage drop compensation v%, the most appropriate parameters of improved energy management strategy and ESS configuration on each substation will be obtained simultaneously by the optimization method based on a genetic algorithm.

4.2. Novel Optimization Method

4.2.1. Genetic Algorithm

The genetic algorithm (GA) is a global optimal searching algorithm based on Darwin's nature evolution theory and Mendel's genetics and mutation theory. It consists of three parts: encoding, fitness evaluation, and genetic manipulation [23,24,25]. Combined with paper demands, the basic procedures of the genetic algorithm are shown as follows.

Encoding

The energy management strategy and configuration of ESSes installed in seven TSSs can be encoded by 16 numbers as shown in Figure 15, where each *X* chromosomere presents a population individual. The first nine numbers represent nine pending parameters of improved energy management strategy, u_1, u_2, $u_3$represent U_{low}, U_{ref1}, U_{ref2}; the last seven numbers represent seven pending numbers of super-capacitor strings installed in seven different traction substations.

$$X = [\underbrace{k_{p1}k_{i1}k_1k_{p2}k_{i2}k_2u_1u_2u_3}_{\text{energy management strategy}}\overbrace{\underbrace{x_1x_2x_3x_4x_5x_6x_7}_{\text{configuration}}}^{TSS\times 7}]$$

Figure 15. Set of *X* chromosomere.

Objective Function ObjV

In this paper, the optimization of energy management strategy and configuration of ESSes is to obtain the maximum objective function *ObjV*, the reciprocal of *ObjV* is the value of fitness. They are calculated as follows:

$$\begin{cases} ObjV[X] = \omega \cdot e\%[X] + (1-\omega) \cdot v\%[X] \\ Fitness[X] = \dfrac{1}{ObjV[X]} \end{cases} \quad (19)$$

where ω is the weight coefficient of economic efficiency e%. *ObjV*[*X*] is the objective function when the energy management strategy, allocation, and size of ESSes are set by *X*.

Genetic Manipulation

Genetic manipulation includes three basic steps—selection, crossover, and mutation. From the view of operators, the genetic algorithm is well-suited to solve combination optimization problems. Compared with other intelligence algorithms, a genetic algorithm has a higher rate of convergence, more efficient calculation, and higher robustness for combination optimization and discrete optimization.

4.2.2. Process of Novel Optimization Method

The schematic diagram of the novel optimization method, which combines a genetic algorithm and simulation platform of urban rail power supply system, is shown as Figure 16. A genetic algorithm can constantly optimize the chromosomere of the population individuals, which means the energy management strategy and configuration of ESSes are optimized constantly. The newfound energy management strategy and configuration of ESSes would be entered into the simulation platform, and obtain their *ObjV*, *e%*, *v%*through the simulation. According to the *ObjV*, *e%*, *v%*, the genetic algorithm can continue the further and cyclic optimization. According to the optimization results by a large number of simulation calculations, the genetic algorithm converges to the global optimum with the increase of evolution generation.

For every different objective function, the genetic algorithm will take 5.5 days to obtain the corresponding optimal solution. Of course, if several workstations work simultaneously, the total simulation time can be effectively decreased. The simulation platform of urban rail power supply system is established by software Matlab 7.10.0(2010a). The hardware performance of our workstations that implement the simulation platform is shown as Table 5. It is worth mentioning that increasing the population size or improving the genetic algorithm by means of a hybrid algorithm can improve the convergence speed and decrease the evolution generation.

The relevant parameters of the genetic algorithm are given by Table 6; *NIND* is population size, *PRECI* is the length of individual, *MAXGEN* is maximum evolution generation, P_c is the crossover rate, P_m is the mutation rate, and *GGAP* is generational gap.

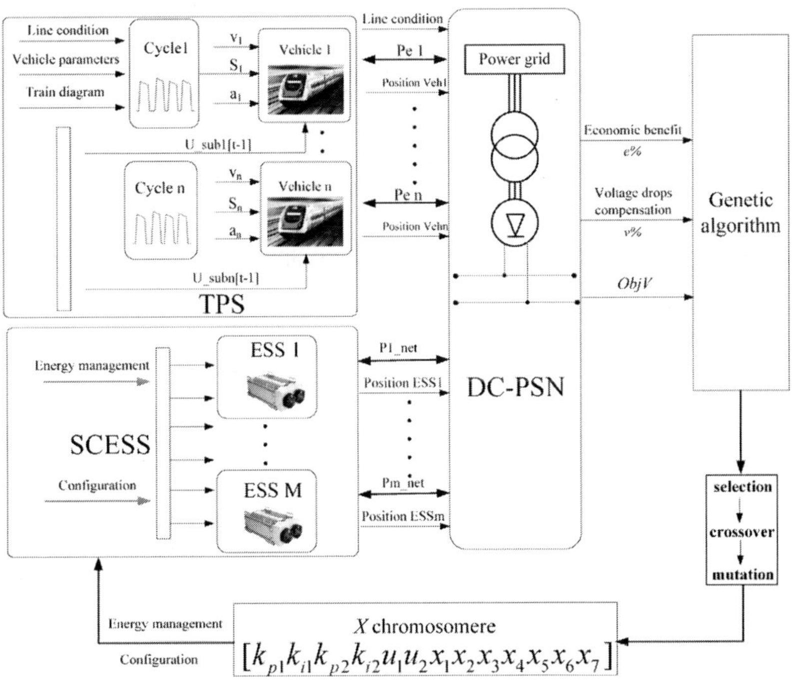

Figure 16. The schematic diagram of novel optimization method.

Table 5. The parameters of hardware platform.

Hardware	Parameter
CPU	Intel(R) Xeon(R) CPU E5649 @ 2.53GHz × 2
RAM	64 GB
GPU	NVIDIA Quadro 4000

Table 6. The parameters of improved genetic algorithm.

NIND	*PRECI*	*MAXGEN*	P_c	P_m	*GGAP*
40	20	100	0.7	0.015	0.95

4.3. Optimization Result Analysis

As shown in Figure 17, the simulation comparison result between two different optimization methods with corresponding optimum *ObjV* are obtained separately under different values of weight coefficient ω. Based on a genetic algorithm, both optimization methods can obtain optimal *ObjV* with an increase of evolution generation, but the novel optimization method can obtain much higher *ObjV*. The values of maximum *ObjV* as well as corresponding economic efficiency e% and voltage drop compensation, v%based on two optimization methods and different values of weight coefficient ω are shown in Table 7. The values of parameters to determine energy management strategy and configuration of ESSes on every substation can be obtained separately based on two optimization methods and different values of weight coefficient ω, as shown in Table 8 and Table 9.

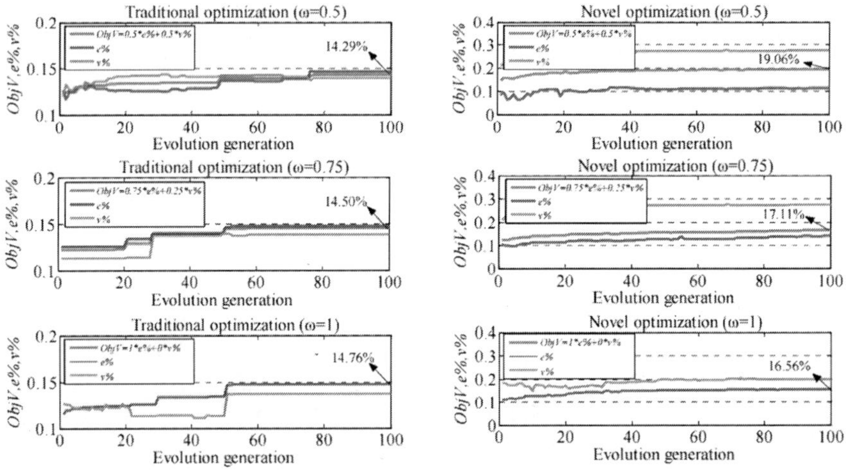

Figure 17. Simulation comparisons of two optimization methods.

Table 7. Maximum *ObjV* obtained by different optimization method.

Optimization Method	ω	Maximum ObjV	Economic Efficiency *e*%	Voltage Compensation Rate *v*%
Traditional optimization	0.5	14.29%	14.65%	13.93%
Traditional optimization	0.75	14.50%	14.72%	13.84%
Traditional optimization	1	14.76%	14.76%	13.70%
Novel optimization	0.5	19.06%	15.06%	23.05%
Novel optimization	0.75	17.11%	15.79%	21.06%
Novel optimization	1	16.56%	16.56%	17.20%

Table 8. The parameters of optimal energy management strategies.

Optimization Method	ω	Energy Management Strategy of ESSes								
		kp1	*ki1*	*k1*	*kp2*	*ki2*	*k2*	U_{low}	U_{ref1}	U_{ref2}
Traditional optimization	-	50	50	-	50	50	-	-	850.0	800.0
Novel optimization	0.5	298	90	0.011	0.158	44.39	0.075	771.0	836.1	802.0
Novel optimization	0.75	193	83	0.006	1.20	40.34	0.037	772.0	836.2	806.4
Novel optimization	1	18	76	0.002	19.33	39.85	0.008	779.1	836.5	811.8

Table 9. Optimized location and size of ESSes.

Optimization Method	ω	TSS No. and Set Numbers of ESSes						
		1	**2**	**3**	**4**	**5**	**6**	**7**
Traditional optimization	0.5	0	16	15	0	10	0	13
Traditional optimization	0.75	0	14	15	0	10	0	14
Traditional optimization	1	0	14	15	0	10	0	13
Novel optimization	0.5	0	18	11	0	17	0	17
Novel optimization	0.75	0	18	10	0	10	0	17
Novel optimization	1	0	14	16	0	8	0	7

From Figure 17 and Table 7, whatever the value of ω, novel optimization method can obtain much higher*ObjV*, *e*% and *v*% compared to traditional optimization method. With the increase of ω from 0.5 to 0.75 to 1, the maximum *ObjV* of ESSes obtained by traditional optimization is 14.29%, 14.50%, and 14.76%, which can be increased to 19.06%, 17.11%, and 16.56%, respectively, by the novel optimization method. And both economic efficiency *e*% and voltage drop compensation *v*% can be improved effectively by the novel optimization compared to traditional optimization. By the novel optimization method, economic efficiency e% can be improved because of more appropriate energy management, less line loss, and voltage drop compensation v% can be improved effectively because of the function of U_{low}.

From Table 8, the adopted energy management strategy of the traditional optimization method is constant, and it is only determined by six parameters. By contrast, the energy management strategy obtained by the novel optimization method is determined by nine parameters. The novel optimization method can optimize the energy management, location, and size of ESSes simultaneously. Under different values of weight coefficient ω, the best energy management strategy is different and among the nine relevant parameters appear some regularities. k_{p1}, k_1, k_{p2}, k_2 are more important factors that affect the performance of the metro system, and k_{i1}, k_{i2}, U_{low}, U_{ref1}, U_{ref2} have smaller changes. Without regard to the integral term, the best energy management strategy of ESSes for different value of weight coefficient ω is shown in Figure 18. For charging energy management strategy, the value of k_{p1} (the slope of charging current *vs.* U_{sub}) and k_1 (the slope of charging current *vs.* U_{sc}) decrease with the increase of weight coefficient ω. For discharging energy management strategy, the value of k_{p2} (the slope of discharging current *vs.* U_{sub}) increases and k_2 (the slope of discharging current *vs.* U_{sc}) decrease with the increase of weight coefficient ω.

Table 9 shows the optimal location and size of ESSes obtained by two different optimization methods. By contrast, two optimization methods ultimately configure super-capacitor ESSes in same location of substations, and the size of ESSes tend to be smaller with the increase of weight coefficient ω. Configuring ESSes in fewer substations with one or two substation spacing and decreasing the size of ESS installed in one substation can reduce the installation cost, but the distance between the train and ESS will also increase, which causes higher line loss and less energy recovered and voltage drop compensation v% will also decrease. The best compromise between economic efficiency e% and voltage drop compensation v% under different value of weight coefficient ω can be obtained by two optimization methods. Compared to the traditional optimization method, the best configuration of ESSes obtained by the novel optimization method changes are more intense and it can achieve much higher *ObjV* under different values of weight coefficient ω.

The maximum *ObjV* with corresponding economic efficiency e% and voltage drop compensation v% for different value of weight coefficient ω can be obtained by novel optimization method as shown in Table 10and Figure 19. From Figure 19, when ω increases from 0.3 to 1, Economic efficiency e% increases from 0.1465 to 0.1656, and voltage drop compensation v% decreases from 0.2335 to 0.1720. According to its own optimization requirement and the concrete result obtained by the novel optimization method in Figure 19, Subway Company could choose the best value of weight coefficient ω for itself.

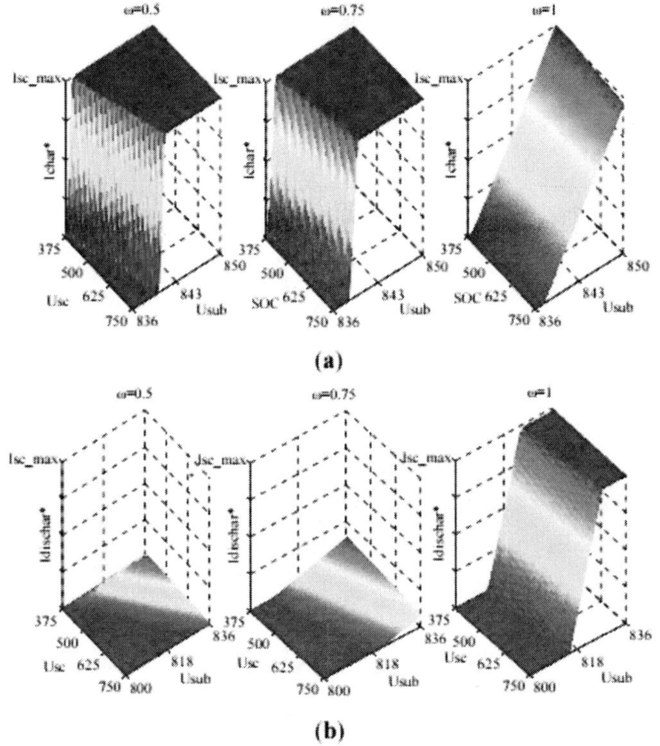

Figure 18. Best energy management strategy of ESSes. (**a**) Charging energy management strategy; and (**b**) discharging energy management strategy.

Table 10. Optimal *ObjV*, *e%* and *v%* obtained by novel optimization method.

ω	0.3	0.4	0.5	0.6	0.7	0.8	0.9	1
ObjV	0.2074	0.1989	0.1906	0.1826	0.1747	0.1685	0.1663	0.1656
e%	0.1465	0.1485	0.1506	0.1509	0.1525	0.1631	0.1654	0.1656
v%	0.2335	0.2325	0.2305	0.2301	0.2265	0.1902	0.1744	0.1720

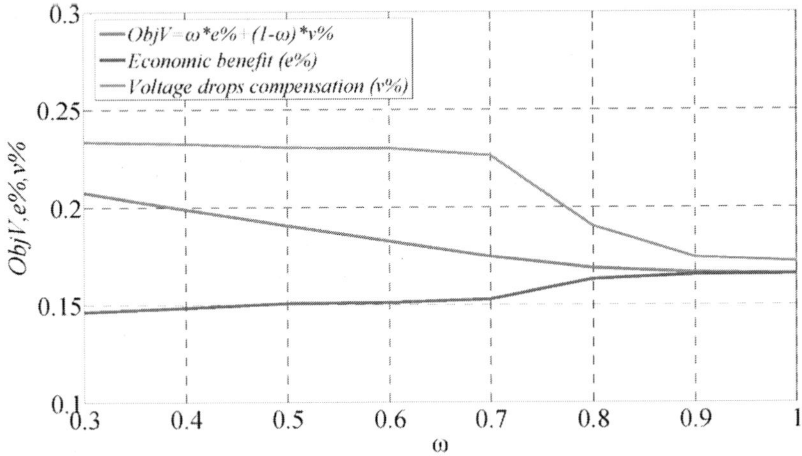

Figure 19. Optimal *ObjV*, *e%* and *v%* obtained by the novel optimization method.

5. CONCLUSIONS

Firstly, this paper establishes the proper simulation platform of a metro system that contains seven substations to simulate the electrical power flow by Matlab/Simulink. Then, two evaluation functions are set up from the perspectives of economic efficiency and voltage drop compensation. Ultimately, a novel optimization method is put forward, which can optimize the energy management strategy, location, and size of ESSes simultaneously by the combination of a genetic algorithm and simulation platform of a metro system. With actual parameters of a Chinese metro line applied in the simulation comparison, the proposed novel optimization method can achieve much better performance of a metro system from the perspectives of*ObjV* and two evaluation functions. The simulation result obtained by the novel optimization method shows that with the increase of weight coefficient ω, the optimal energy management strategy is different and the nine relevant parameters appear with some regularities, among them k_{p1}, k_1, k_{p2}, and k_2 are more important factors that affect the performance of the metro system. Additionally, novel optimization methods can also optimize the configuration of ESSes, which can achieve the best compromise between economic efficiency e% and voltage drop compensation v%. The novel optimization method and its optimized result can provide valuable reference to Subway Company.

ACKNOWLEDGMENTS

This research was supported by the Major State Basic Research Development Program of China (973 Program: 2011CB711100) and I13L00100 from Beijing Laboratory of Urban Rail Transit.

AUTHOR CONTRIBUTIONS

Zhongping Yang contributed to the conception of the study, Huan Xia and Huaixin Chen contributed significantly to analysis and manuscript preparation, Fei Lin helped perform the analysis with constructive discussions and Bin Wang provided the line and vehicle data.

CONFLICTS OF INTEREST

The authors declare no conflict of interest.

REFERENCES

1. Barrero, R.; van Mierlo, J.; Tackoen, X. Energy savings in public transport. *IEEE Veh. Technol. Mag.* **2008**, *3*, 26–36.
2. Barrero, R.; Tackoen, X.; Van Mierlo, J. Improving energy efficiency in public transport: Stationary supercapacitor based energy storage systems for a metro network. In Proceedings of the IEEE Vehicle Power and Propulsion Conference (VPPC'08), Harbin, China, 3–5 September 2008; pp. 1–8.
3. Ogasa, M. Energy saving and environmental measures in railway technologies: Example with hybrid electric railway vehicles. *IEEJ Trans. Electr. Electron. Eng.* **2008**, *3*, 15–20.
4. Yamanoi, T.; Kawahara, K.; Kani, Y.; Kodama, Y. Measurement of utilized regenerative power in DC feeding system. In Proceedings of the 2013 IEEJ Industry Applications Society Conference, Yamaguchi, Japan, 28–30 August 2013.

5. Avanzo, S.D.; Iannuzzi, D.; Murolo, F.; Rizzo, R.; Tricoli, P. A sample application of supercapacitor storage systems for suburban transit. In Proceedings of the Electrical Systems for Aircraft, Railway and Ship Propulsion (ESARS), Bologna, Italy, 19–21 October 2010; pp. 1–7.

6. Ibrahim, H.; Ilinca, A.; Perron, J. Energy storage systems—Characteristics and comparisons. *Renew. Sustain. Energy Rev.* **2008**, *12*, 1221–1250.

7. Gabash, A.; Li, P. Active-reactive optimal power flow in distribution networks with embedded generation and battery storage. *IEEE Trans. Power Syst.* **2012**, *27*, 2026–2035.

8. Gabash, A.; Li, P. Flexible optimal operation of battery storage systems for energy supply networks.*IEEE Trans. Power Syst.* **2013**, *28*, 2788–2797.

9. Grbovi, P.J.; Delarue, P.; Le Moigne, P.; Bartholomeus, P. The ultracapacitor-based controlled electric drives with braking and ride-through capability: Overview and analysis. *IEEE Trans. Ind. Electron.***2011**, *58*, 925–936.

10. Iannuzzi, D. Improvement of the energy recovery of traction electrical drives using supercapacitors. In Proceeding of 13th Power Electronics and Motion Control Conference (EPE-PEMC 2008), Poznan, Poland, 1–3 September 2008; pp. 1469–1474.

11. Ciccarelli, F.; Iannuzzi, D.; Tricoli, P. Control of metro-trains equipped with onboard supercapacitors for energy saving and reduction of power peak demand. *Transp. Res. C Emerg. Technol.* **2012**, *24*, 36–49.

12. Ciccarelli, F.; Del Pizzo, A.; Iannuzzi, D. Improvement of energy efficiency in light railway vehicles based on power management control of wayside lithium-ion capacitor storage. *IEEE Trans. Power Electron.* **2014**, *29*, 275–286.

13. Ciccarelli, F.; Iannuzzi, D.; Spina, I. Comparison of energy management control strategy based on wayside ESS for LRV application. In Proceedings of the 39th IEEE Annual Conference on Industrial Electronics Society (IECON 2013), Vienna, Austria, 10–13 November 2013; pp. 1548–1554.

14. Battistelli, L.; Fantauzzi, M.; Iannuzzi, D.; Lauria, D. Energy management of electrified mass transit systems with energy storage devices. In Proceedings of the 2012 International

Symposium on Power Electronics, Electrical Drives, Automation and Motion (SPEEDAM), Sorrento, Italy, 20–22 June 2012; pp. 1172–1177.

15. Iannuzzi, D.; Pighetti, P.; Tricoli, P. A study on stationary supercapacitor sets for voltage droops compensation of streetcar feeder lines. In Proceedings of the 2010 Electrical Systems for Aircraft, Railway and Ship Propulsion (ESARS), Bologna, Italy, 19–21 October 2010; pp. 1–8.

16. Wang, B.; Yang, Z.; Lin, F.; Zhao, W. An improved genetic algorithm for optimal stationary energy storage system locating and sizing. *Energies* **2014**, *7*, 6434–6458.

17. Iannuzzi, D.; Ciccarelli, F.; Lauria, D. Stationary ultracapacitors storage device for improving energy saving and voltage profile of light transportation networks. *Transp. Res. C Emerg. Technol.* **2012**, *21*, 321–337.

18. Radcliffe, P.; Wallace, J.S.; Shu, L.H. Stationary applications of energy storage technologies for transit systems. In Proceedings of the 2010 IEEE Electric Power and Energy Conference (EPEC), Halifax, NS, Canada, 25–27 August 2010; pp. 1–7.

19. Barrero, R.; Tackoen, X.; Van Mierlo, J. Stationary or onboard energy storage systems for energy consumption reduction in a metro network. *Proc. Inst. Mech. Eng. F J. Rail Rapid Transit* **2010**, *224*, 207–225.

20. Teymourfar, R.; Asaei, B.; Iman-Eini, H. Stationary super-capacitor energy storage system to save regenerative braking energy in a metro line. *Energy Convers. Manag.* **2012**, *56*, 206–214.

21. Iannuzzi, D.; Pagano, E.; Tricoli, P. The use of energy storage systems for supporting the voltage needs of urban and suburban railway contact lines. *Energies* **2013**, *6*, 1802–1820.

22. Battistelli, L.; Ciccarelli, F.; Lauria, D.; Proto, D. Optimal design of DC electrified railway stationary storage system. In Proceedings of the 2009 International Conference on Clean Electrical Power, Capri, Italy, 9–11 June 2009; pp. 739–745.

23. Guo, P.; Wang, X.; Han, Y. The enhanced genetic algorithms for the optimization design. In Proceedings of the 3rd International Conference on Biomedical Engineering and Informatics (BMEI

2010), Yantai, China, 16–18 October 2010; Volume 7, pp. 2990–2994.

24. Tegani, I.; Aboubou, A.; Becherif, M.; Ayad, M.Y.; Kraa, O.; Bahri, M.; Akhrif, O. Optimal sizing study of hybrid wind/PV/diesel power generation unit using genetic algorithm. In Proceedings of the 4th International Conference on Power Engineering, Energy and Electrical Drives (POWERENG 2013), Istanbul, Turkey, 13–17 May 2013; pp. 134–140.

25. Jiao, L.Y.; Lei, H.Z. The Application of genetic algorithm in fitting the spatial variogram. In Proceedings of the 2011 International Conference on Computer Science and Network Technology (ICCSNT), Harbin, China, 24–26 December 2011.

4

CHAPTER

PARAMETER OPTIMIZATION FOR LOCAL POLYNOMIAL APPROXIMATION BASED INTERSECTION CONFIDENCE INTERVAL FILTER USING GENETIC ALGORITHM: AN APPLICATION FOR BRAIN MRI IMAGE DE-NOISING

Nilanjan Dey et al.

[1]*Department of Information Technology, Techno India College of Technology, Kolkata 700156, India*

Abstract

Magnetic resonance imaging (MRI) is extensively exploited for more accurate pathological changes as well as diagnosis. Conversely, MRI suffers from various shortcomings such as ambient noise from the environment, acquisition noise from the equipment, the presence of background tissue, breathing motion, body fat, *etc.* Consequently, noise reduction is critical as diverse types of the generated noise limit the efficiency of the medical image diagnosis. Local polynomial approximation based intersection confidence interval (LPA-ICI) filter is one of the effective de-noising filters. This filter requires an adjustment of the ICI parameters for efficient window size selection. From the wide range of ICI parametric values, finding out the best set of tunes values is itself an optimization problem. The present study proposed a novel technique for parameter optimization of LPA-ICI filter using genetic algorithm (GA) for brain MR images de-noising. The experimental results proved that the proposed method outperforms the LPA-ICI method for de-noising in terms of various performance metrics for different noise variance levels. Obtained results reports that the ICI parameter values depend on the noise variance and the concerned under test image.

Keywords: magnetic resonance imaging; image de-noising; local polynomial approximation filter; optimization algorithms; genetic algorithm

1. INTRODUCTION

Brain MR Image analysis is a common clinical activity which is used to diagnose many neurological diseases such as Alzheimer's disease, multiple sclerosis, brain tumor, *etc.* A computer-aided system helps automating this clinical activity in order to obtain faster and reliable analysis. Recently, an imperative research is focused on the medical image processing in various fields such as [1,2,3,4].

Generally, Magnetic resonance imaging (MRI) is one of the most noninvasive medical imaging powerful modalities endow with highly detailed images of organs and tissues in the human body. It produces detailed images of organs, bone, soft tissues, brain, and all other internal body structures. MRI facilitates the discovery of abnormalities which might be obscured by bone with other imaging methods. However, these MRI images are undergoing numerous noise sources, such as a transmission medium, recording medium (digital sensor, film) and measurement/ quantization errors [5], ambient noise from the environment, acquisition noise from the equipment, breathing motion, body fat, RF coil, RF pulses, the presence of background tissue, field strength, receiver bandwidth, *etc.*

Consequently, the MRI acquired directly from the instruments is commonly inadequate for medical analysis, especially in brain and cardiac [6] because of the existence of high levels of Rician noise [7,8] that degrade the image quality. As a corollary, to achieve consistent analysis, removing MRI image noises is compulsory before conducting any supplementary image processing such as registration, segmentation, classification, and visualization.

Noise modeling is significantly influenced by the capturing instruments. Diverse algorithms are employed based on the noise model. For supplementary analysis, pre-processing and appropriate diagnosis necessitates the development of de-noising methods to refine the image quality and increase the Signal-to-Noise Ratio (SNR).

Recently, removing the noise from images to restore the original image is considered as a serious challenge that has a sturdy research [9,10]. Enormous techniques using variety algorithms used for MRI medical images de-noising, some of them are as follows.

Wavelet methodology has enormous uses in diverse areas as presented in [11,12], as well as it proved its efficiency in de-noising of MRI images as follows. Healy and Weaver [13], dealt with soft-thresholding based on wavelet techniques for de-noising MR images. While, Nowak [14], studied that the square magnitude MRI image, contains a Rician noise model in the threshold-based wavelet de-noising approach to correct the bias produced by the noise.

Through the year 2003 [15], the authors suggested MRI image de-noising technique using adaptive wavelet thresholding. To amplify the significant features, this method multiplied the adjacent wavelet sub bands, and then applied threshold to the multi-scale products to differentiate edge structures in an improved approach from noise. Results of MRI images proved that this method achieved high mean-to-standard-deviation ratio (MSR) and contrast to noise ratio (CNR). During the same year, Jiang *et al.* [16] was presented a background-noise modeling using a Rician distribution. Then, the authors employed the parameter estimation via a Maximum-Likelihood method. Both simulated and real MRI image results proved that the proposed method outperformed other de-noising approaches. Based on spectral subtraction for noise removal in functional magnetic resonance imaging (fMRI) data in [17], the author presented an adaptive signal-preserving technique. The introduced technique had estimated a parametric model for the power spectrum of the random noise from the obtained data derived from the characteristics of the Rician statistical model. The results demonstrated the potential of the suggested method in noise suppression. Meanwhile, a de-noising algorithm for medical images had been proposed in [18], based on a combination of both the wavelet and the total variation minimization methods. In real noisy medical images, results showed that the proposed algorithm offers effective noise removal. As a supplement, this algorithm permitted the implementation of an effective automatic stopping time criterion extensively.

Manjon *et al.* [19] employed spatial averaging for similar pixels. The authors utilized information from all available image components to perform the de-noising process. Later, Rajeesh *et al.* [20] used wave atom shrinkage for MRI images de-noising. The authors established that curvelet shrinkages were better than the wavelet method, as the proposed approach had achieved better SNR. Furthermore, the effectiveness of the proposed method was accentuated with large dataset of real-time normal and pathological MR images.

A review of the noise reduction algorithms for brain MRI images was presented in [21]. The anisotropic diffusion filtering reduces the noise as well as erased the small feature due to its edge enhancement. Another filter type using the spectral subtraction method was performed after acquiring signals from each RF coil separately with less acquisition time [22]. The results showed enhanced signal to noise ratio (SNR) up to 40% in the MRI reconstructed signal.

Meanwhile, the authors in [23] implemented a de-noising algorithm on MR brain images. The proposed de-noising algorithm was concerned with the poor quality images manipulation as well as to improve the deteriorated SNR. The noise was removed from the MRI images using the median filter, the weighted median and the adaptive filter. Lastly, the evaluation performance of these filters showed that using the weighted median filter was outperforming the other filters with peak signal to noise ratio PSNR = 0.924.

Iftikhar *et al.* [24] proposed a novel improved adaptive nonlocal means (IANLM) algorithm. The authors employed a window adaptation test based on robust threshold criterion that used to select the appropriate pixels for the restoration process. The experimental results were performed at various noise levels of simulated real brain MR images. It is established that the proposed algorithm was better for de-noising jointly with its efficient computation due to proposed modifications. In the year 2014 [25], the authors suggested an enhanced non-local means (ENLM) algorithm as an extension for the previously mentioned IANLM algorithm for brain MRI images with Rician noise. Results proved the superiority of the ENLM compared to other de-noising methods. By adapting the IANLM method [26], an extended nonlocal means (XNLM) algorithm was used to Rician noise in MR images. Furthermore, a wavelet coefficient mixing process was used in the XNLM to mix the wavelet sub-bands of two IANLM-filtered images. Lastly, the XNLM consists of new parameter-free pixel pre-selection method which improved the algorithm computational efficiency.

Most of these algorithms used low-pass filtering and averaging to achieve de-noising, where high frequency coefficients can be removed by filtering. At the same time, since edges also have high frequency components, thus during removing the noise, high frequency components of the edges are also removed. Therefore, using weighted median filters conserved edges while removed noise as proved. Consequently, Klepaczko in 2013 [27], proposed a novel scheme based on the local polynomial approximation (LPA) filter for de-noising to design GPU hardware. Relevance of LPA to high-resolution 3D MR images occurs inefficient. Thus, the author proposed parallelized GPU-accelerated implementation of LPA to remove the noise from the MR data effectively and efficiently. The author concluded that the emphasized designed GPU-based performance of the LPA algorithm was achieved very powerfully. Accomplished within a time frame reduced to milliseconds, generation of LPA masks, no longer entails any significant computational load to Shape-Adaptive Discrete Cosine Transform (SADCT) based noise removal method.

As a consequence, the LPA is efficient for MRI brain images de-noising as it is a non-parametric algorithm. It was implemented in various image processing applications [28,29].The key parameter of the LPA is the choice of its window size (scale), where a signal is convolved with a kernel function of a known form to estimate the values in the locality of a given data point. For adjusting the window size to be adapted to the local image contents, an intersection of confidence intervals (ICI) rule can be used. This LPA-ICI algorithm for image de-noising has various parameters to be adjusted to determine the optimal window size. One of these parameters is Gamma Γ, which has an infinite range. Besides, other parameters are known to be adjusted. Thus, the key idea of the proposed work is to employ an optimization procedure that can be used to adjust the LPA-ICI parameters for each MRI brain images.

Recently, a variety of optimization algorithms have been used in the medical imaging area such as in [30,31,32,33,34,35]. Consequently, Misra *et al.* proposed a method where the noise-reduction filtering is realized by Genetic Algorithm [36]. The genetic operator combines the crossover and adaptive mutation to improve the convergence rate as well as the solution quality of the GA. The suggested technique successfully diminishes the standard deviation and considerably lowers the rectified noise. Meanwhile, by the year 2014, Liu *et al.* conducted a medical image de-noising method using GA based on wavelet threshold [37]. To obtain the optimal threshold, the authors used a GA combined with clustering histogram of the image information. Experimental results proved the improved visual effect of medical image after de-noising and the improved peak signal to noise ratio when using the ordinary wavelet threshold demising method.

Consequently, this work will merge both the Genetic Algorithm optimization technique to support LPA-ICI de-noising filter (LPA-ICI-GA) by selecting the optimized ICI parameters for Rician noise removal. Where, genetic procedures including reproduction, crossover, and mutation are applied to increase the signal-to-noise ratio and minimize the overall computational complexity through using the mean square error (MSE) as the cost function. The main advantage of the GA compared to other optimization algorithms is its robustness. Jointly with, it requires less number of iterations for convergence [38]. Therefore, the contribution of this work is the optimization based GA that used to select the optimal parameters of the LPA-ICI filter. In addition, the supposed benefit of this research is to obtain better PSNR and system performance compared to the LPA-ICI itself.

The remaining sections are organized as follows. Section 2 introduces the Rician distributed noise model and the MR signals de-noising. Then, Section 3 presents the methodology that includes the image de-noising using LPA-ICI and the genetic algorithm for optimization. Section 4 presented the proposed system. Consequently, the simulation results and discussion are conducted through Section 5. Finally, the conclusion and future work are offered in Section 6.

2. RICIAN DISTRIBUTED NOISE AND MR SIGNALS DE-NOISING

MR images are commonly corrupted with Rician noise [39,40]. This type of noise typically produces hazards in low signal-to-noise ratio (SNR) regions, high resolution [41], which enforces a trade-off between SNR and image resolution [42]. Consequently, post-processing noise reduction is employed to achieve the desired MRI image quality. The foremost MRI image noise source is the thermal noise in the patient [43].

Commonly, the MRI image is reformed by computing the inverse discrete Fourier transform of the raw data [44]. The signal component that exists in both real and imaginary orthogonal are affected by additive white Gaussian noise. The noise in the reconstructed complex-valued data is complex white Gaussian noise. Generally, the reconstructed MRI image magnitude is used for automatic computer analysis and visual inspection. Since, the magnitude of the MRI signal is the square root of the sum of the squares of two independent Gaussian variables; it follows a Rician distribution [45]. In low intensity (dark) regions of the magnitude image, the Rician distribution tends to a Rayleigh distribution [46] and in high intensity (bright) regions it tends to a Gaussian distribution. This leads to reduced image contrast where noise increases the mean value of the pixel intensities in dark image regions.

2.1. Rician Noise Mathematical Model

Brain MRI is the procedure of detecting most brain disorders. It affords clear images of the posterior brain and brainstem, which are complex to view on a CT scan. The probability density function (PDF) of the noisy signal that corrupts the brain images has a Rice Distribution. Rician noise in the magnitude MRI occurs from complex Gaussian noise in the original frequency domain measurements.

Assume an original signal *z* is known to be corrupted with Rician distributed noise, so the resulting corrupted noisy image intensity *N* follows a Rician probability distribution, which is given by [47]:

$$P(z,N,\sigma)=\frac{N}{\sigma^2}e\left(-\frac{z^2+N^2}{2\sigma^2}\right)I_0\left(\frac{zN}{\sigma^2}\right) \quad (1)$$

where z, *N*, σ > 0, σ is denoted the noise standard deviation and Io is indicating the modified Bessel function of order zero. Additionally, zσ expressed the SNR, where for high SNR, the Rician distribution approaches a Gaussian, while as the SNR approaches 0 (low), the Rician distribution befalls the Rayleigh distributed. The Rician noise engenders random variation as well as sets up a signal-dependent bias to the data which leads to decreasing the image contrast. Rician noise is not additive as it is dependent on the data itself. Consequently, to add Rician noise to the data, it is essential to construct the data as Rician distributed.

2.2. MR Images De-noising

Generally, image processing adapts images to improve them for feature extraction and to change their structure [48,49]. The effort of image de-noising is to improve an image that is cleaner than its noisy observation [50]. Filtering is considered as the most elementary operation in many biomedical image processing applications, where it reduces the noise level and improves the image quality. In general, the selection of a suitable de-noising algorithm is dependent on the specific targeted application. De-noising algorithms can be classified into two basic classes, namely spatial filtering methods (linear and nonlinear), and transform domain filtering methods. The latter category can be subdivided along with the basic function selection to adaptive/ data non-adaptive approaches. Wavelet filtering is considered as non-data adaptive approach. In the last two decades, the wavelet transform filtering [51] has an increasing esteem in various wavelet-domain de-noising algorithms, especially with MRI images [52,53,54]. In addition, diverse approaches to identify the threshold values of wavelet coefficients were conducted. Dissimilar the wavelet-domain filtering, a spatial adaptive filter whose filter support size can be selected by the relative intersection of confidence intervals is the local polynomial approximation filter [55,56]. Consequently, much consideration in wavelet domain filtering was given to other numerous de-noising approaches such as anisotropic diffusion [57], nonnegative sparse coding [58], hidden Markov models [59], Wave atom transform [60], Local polynomial approximation filter, *etc.*

3. METHODOLOGY

The local polynomial approximation (LPA) filter received extensive attention, as it is an efficient and flexible spatial adaptation and simple implementation. Through LPA, each sample is modeled in the MR image as a local polynomial with a window size (kernel) having a certain bandwidth matrix rather than filtering the whole image. Then, resolve a series of local least-squares (LS) estimation problem through neighboring samples from the images. Since, images may vary significantly; it is very decisive to select a proper local window size to accomplish the best bias-variance tradeoff in estimating the LPA coefficients. Typically, the optimal local window must be selected to minimize the mean squared error (MSE).

Consequently, in this paper, an innovative MR brain image de-noising method using the LPA-ICI filter based on genetic algorithm is proposed. Where, the genetic algorithm is used to obtain the optimal parameters used for the window size (scale) of the LPA-ICI according to the brain MRI image under concern.

3.1. Image De-Noising using LPA-ICI

The Local polynomial approximation approach is utilized for nonparametric estimation by polynomial data fit in a sliding window with fixed size [61]. Lately, Katkovnik *et al.* [62] proposed an intersection confidence intervals (ICI) rule to support the LPA algorithm for local adaptive scale selection. For a given set of scale parameters in arising order, the ICI rule can define a near optimal scale by comparing the estimated confidence intervals with diverse scales from the scale set. However, the ICI method can only select the scale from the finite scale set. Therefore, a nonlinear powerful LPA-ICI filter which is adapted to the unknown smoothness of the signal is produced.

Assume that an M1×M2 image can be viewed as the observations of a 2D function. If the noisy observations of the image intensity Z(x) on the grid of argument values Y(s)=(y1(s), y2(s)) being two dimensional vectors with the components y1(s) and y2(s). Where, *S* indicates the corresponding s^{th}pixel of the image.

Accordingly, the noisy image can be represented as [63,64]:

$$N(x) = Z(x) + \eta(x) \tag{2}$$

where η(x) is independent and identically distributed random Gaussian errors. For de-noising, the main objective is to estimate Z(x) depending on the observation N(x) with mean square error (MSE) as small as possible. To apply the LPA, a part of the truncated Taylor series is utilized to approximate the varying intensit Z(x).

Then, locally this expansion is exploited in a relatively small area to calculate the estimate for a single "central" pixel only. For the next pixel, the calculations are repeated. Consequently, this point-wise procedure establishes a nonparametric feature of the estimation. The LPA function is given by:

$$J_h(x) = \sum w_h(y(s)-x)\left(N(y(s)) - C^T\varphi(y(s)-x)\right)^2 \tag{3}$$

$$\varphi(y) = (\varphi_1(y), \varphi_2(y), \dots, \varphi_U(y))^T \tag{4}$$

where, $\varphi(y)$ is a set of linear independent 2D polynomials of the powers from 0 to u with ϕ=1.

The window function with window size *h* (scale) is defined as:

$$w_h(x) = (\frac{x}{h})w(\frac{x}{h}) \tag{5}$$

This function introduces the fitting localization with respect to the center *x*. The estimate with respect to the polynomial components of *Z*(*x*) can be represented as:

$$\hat{Z}(x,h) = \sum_s g_1(x, y(s), h) N(y(s)) \tag{6}$$

$$g = \Phi^{-1} w_h(y(s)-x)\varphi(y(s)-x) \tag{7}$$

$$\Phi = \sum_s w_h(y(s)-x)\varphi(y(s)-x)\varphi^T(y(s)-x) \tag{8}$$

where, g_1 is the first element of the vector *g*.

The linear transform in equation (6) can be applied to the data. Since, the window size selection is a crucial point of the efficiency of the local estimators. When *h* is relatively small, the LPA offers a good approximation of *y(x)*, but useless data. Hence, the estimates are more variable and sensitive with respect to the noise. The best choice of *h* involves a trade-off between the bias and variance, which depends on the order of the derivatives being involved in the LPA, the noise variance, a sample period, and values of the derivatives of *y(x)* beyond the order used in the LPA. The estimation error can be represented as:

$$e(x,h) = z(x) - \hat{z}(x,h) \tag{9}$$

From further derivatives in [65], it is shown that the ideal window size h∗(x) depends on the polynomial degree and the ideal bias-to-variance tradeoff. This optimal window size is achieved when the ratio between the bias and standard deviation is equal to a constant Γ.

For capable de-noising (filtering), the window size is a significant parameter, where for less noise value, the scale should be selected as small as possible (smaller bias). While as the noise value increases, the scale should be increased in order to suppress noise effects. So, the LPA with a fixed window width requires adapted window size. This is achieved using the LPA-ICI algorithm combines adaptive window that supports the LPA algorithm. The adaptive-window selection procedure is based on the approximate minimization of the mean squared estimation error using the bias-to-variance tradeoff approach. Therefore, using the ICI rule enhances the filter performance by selecting the appropriate window length according to the used image. For more details on the LPA-ICI algorithm for near optimal window size selection from a set of windows, the reader can refer to [56]. According to the ICI, the optimal window h∗ is the largest one followed by empty confidence interval, as no intersection between the intervals will exist at the optimal window width.

The interchangeable window size is the key parameter of the LPA-ICI filter. The steps of the ICI scale-adaptation are:

Step 1: **Set** Γ = Γl, x=y(s) and s=1,...,E.

Step 2: **Calculates** the estimates z^(x,h), the adaptive window size h∗(x) and the estimates z^(x,h∗)for h=hi, i=1,..,J. (Calculate for different scales the estimates and compare them)

Step 3: **Repeat** steps 1 and 2 for all y(s).

Step 4: **Find** Γ^ and select the estimates z^(x,h∗) corresponding to Γ^ as the final ones. (Find the adaptive scale which is the largest where the estimate does not vary significantly from the estimates corresponding to the smaller scales).

Therefore, the LPA-ICI algorithm searches for the largest local neighborhood of the point of estimation where the LPA assumptions fit well to the data. Where, the ICI rule defines the adaptive scale for each point pixel of the image. Thus, point-wise adaptive signal and image processing is achieved.

The resulting adaptive estimator is nonlinear even for the linear LPA as the nonlinearity of the method is incorporated in the ICI rule itself.

From the previous algorithm, it is clear that, the parameter Γ plays a key role in the appropriate LPA-ICI filter support selection. Too large Γ outcomes in signal over-smoothing, while too small Γ results in signal under-smoothing.

So, the ICI rule requires threshold parameter adjustment for each image and noise level. As a result, the foremost drawback of the ICI rule is its dependence on the threshold parameter Γ to determine the optimal window size. Also, the exact optimal window size is not included in the windows set. Where, a point the locally adaptive scale parameter h will control the smoothing amount to be accomplished at that point. When h a small, image detail for example is the edges will be preserved. Conversely, probable additive noise removal can't be done effectively. On the contrary, a large-scale has superior de-noising properties with the cost of possibly blurring of the image details. Therefore, a locally adaptive scale selection method is essential to the LPA in image processing.

Asymptotically, these adaptive estimators allow getting a near-optimal quality of the signal recovery. Therefore, to obtain the optimal parameter selection an optimization algorithm can be used as in the proposed work.

3.2. Optimal Selection of Parameters using Genetic Algorithm

Rician noise generates random variation in the data and has an influence on the MR images that reduces image contrast. The LPA-ICI algorithm engages a few parameters which should be tuned for a brain MRI images de-noising in order to obtain optimized performance. In this proposed work, the GA is chosen for optimization. It is considered a search process to facilitate optimization of the structures it generates. Genetic algorithm is characterized by its robustness and high efficiency for complex search problems without being stuck in local extreme. It is known as a heuristic algorithm which is efficient to reach optimal or near-optimal global solution. It uses a fitness function that determines the quality of an individual in the population.

In a GA, a population of individuals illustrated by some chromosomes is iteratively updated by concerning imperative operators of selection, mutation and crossover to solve the problem. Each individual is evaluated by a fitness function that controls the population evolution in order to optimize it.GA can be used for any kind of continuous or discrete optimization problem. The optimization is achieved using natural swap of genetic material between parents. Offspring's are created from parent genes. Reproduction creates offspring and has a combination of features inherited from each parent. The fitness of offspring's is then measured and only the fittest offspring is granted to breed. Where, the fitness values are used in a natural selection process to select which potential solutions will persist on to the next generation, and which will die out.

Synthetically, genetic material is replaced by strings of bits and natural selection substituted by the fitness function. Mating of parents is represented by crossover and mutation operations. GA is a parallel optimization technique that relies on the basics of evolution for optimizing a group of solutions at once. The general steps for the GA can be illustrated as [66]:

Step 1: **(Begin)** Generate random population of chromosomes. (Suitable solutions for the problem)

Step 2: **(Evaluate population-Fitness)** In the population, evaluate the fitness of each chromosome.

Step 3: (**New population**) Create a new population. (By repeating the following steps until the new population is complete)

a) **Selection**: From a population, select two parent chromosomes according to their fitness. (Better fitness, provides bigger chance to be selected to be theparent).
b) **Crossover**: With a crossover probability, cross over the parents to form new offspring (children). If no crossover was performed, offspring is the exact copy of parents.
c) **Mutation**: With a mutation probability, mutate new offspring at each locus.
d) **Accepting**: Place new offspring in the new population.

Step 4: (**Replace**) Use new generated population for a further algorithm run.

Step 5: **(Test)** If the end condition is satisfied, stop, and return the best solution in current population.

Step 6: (**Loop**) Go to step 2.

The genetic algorithm performance is largely influenced by crossover and mutation operators.As a conclusion, the most significant characteristics for the GA are the Initial population (possible solutions), the fitness function, and the Genetic operators.

4. THE PROPOSED LPA-GA SYSTEM

Since, GA proves its efficiency for better convergence towards global optimization than any other global search algorithm. Consequently, the proposed method in this work makes use of genetic algorithm to determine the optimal parameter values in the LPA-ICI de-noising filter in order to reduce Rician noise by optimizing the mean square error taken as a cost function.

The brain tumor is a defined as a group of abnormal cells that grows inside of the brain or around the brain. Tumors can directly destroy all healthy brain cells. As well as it can indirectly damage healthy cells. The MRI is the main modality that concerned with brain images and other body parts. Thus, the brain images are used to test the proposed novel algorithm as the brain images considered as the second leading cause of cancer death.

The LPA-ICI-GA based filtering procedure maintains key image element and characteristics, reduces the Rician noise. In this paper, simulated Rician noise introduced in the test image slice and then the noise is filtered out using LPA-ICI based GA optimization.

Figure 1 depicts the proposed system using the following steps:

Step 1. The original MRI-brain image is added with Rician noise with various values of “s” ranging from 0.1 to 0.6 to produce the corrupted images.

Step 2. The genetic algorithm is used to optimize the input parameters of LPA-ICI.

Step 3. The input parameters include the Gamma ICI parameter Γ, sharpness parameter, number of directions = 4 or 8, and fusion parameter which is classical or piecewise.

Step 4. The upper and lower bounds of the input parameters are set as [−1 to 10] for the Gamma ICI parameter, [0–5] for sharpness parameter, [[4 or 8] for direction parameter, [[1 or 2] for fusion parameter.

For each solution, the fitness function is to be computed based on the obtained PSNR and mean structural similarity (MSSIM) values. The solution set with the best fitness value is then selected and stored as the best solution for the gained parameters. The fitness function (FF) is calculated between the original and restored images, can be expressed as follows:

$$FF = PSNR + \Omega \tag{10}$$

where, Ω is the correlation factor defined as (Ω = 100* MSSIM). Here, the normal values of the MSSIM fall in the range 0~1; consequently, MSSIM are multiplied by 100, as the PSNR values may reach the value of 100.

Step 5. The fitness function uses the LPA-ICI with optimized parameters given by the GA to produce the restored image.

Step 6. The genetic algorithm is used to maximize the fitness function (by multiplying the result by −1). The number of generations is set to 100 by default.

Step 7. The genetic algorithm stopped around 50 generations for all noise ratios. Where, convergence occurred around 25 generations.

Step 8. The optimized parameters generated by the genetic algorithm are then given to the LPA-ICI filter to de-noise the corrupted image.

Step 9. The fitness function (FF), calculates PSNR+100*MSSIM between the original image and the restored image. Where, the LPA-ICI with parameters selected by GA is used to produce the restored image.

Through the GA stage, evaluation is executed via the fitness function which depends on the precise problem (de-noising) and is the optimization objective of the GA.

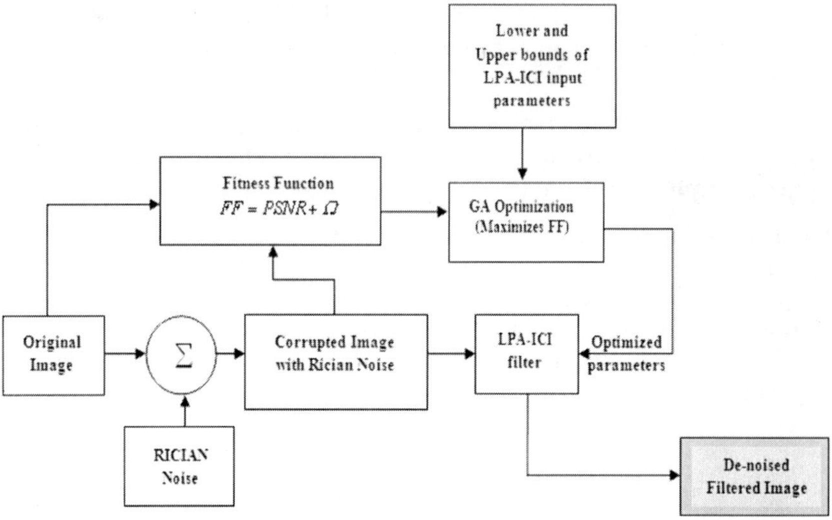

Figure 1. Proposed system block diagram.

5. RESULT AND DISCUSSION

To study the local polynomial approximation based intersection confidence interval is supported by genetic algorithm for MRI brain images de-noising. The Matlab R2008a (Version 7.6.0.324) software is used. The algorithm has been executed at a base station in a Windows 7 platform having 6GB RAM, Intel Core i5-2410M CPU 2.30GHz processor. The experimental results are obtained using the public MRI data set of 27 horizontal slices of brain images, available at [67].

5.1. Comparing the results of LPA-ICI-GA versus LPA-ICI Algorithm

Figure 2 illustrates the results obtained from employing the LPA-ICI and the LPA-ICI-GA algorithms for de-noising the MRI brain slice images that corrupted with Rician noise of (s = 0.1) variance value. The figure demonstrates that the proposed system is able to restore the MRI image efficiently even with the existence of the Rician noise. Also, the image restored by LPA-ICI-GA is better than the restoration by LPA-ICI. It is clear that still some white dots are there when using LPA-ICI algorithm in Figure 2c. These results are shown when s = 0.1 where the proposed system provides perfect results for all noise variances less than 0.1.

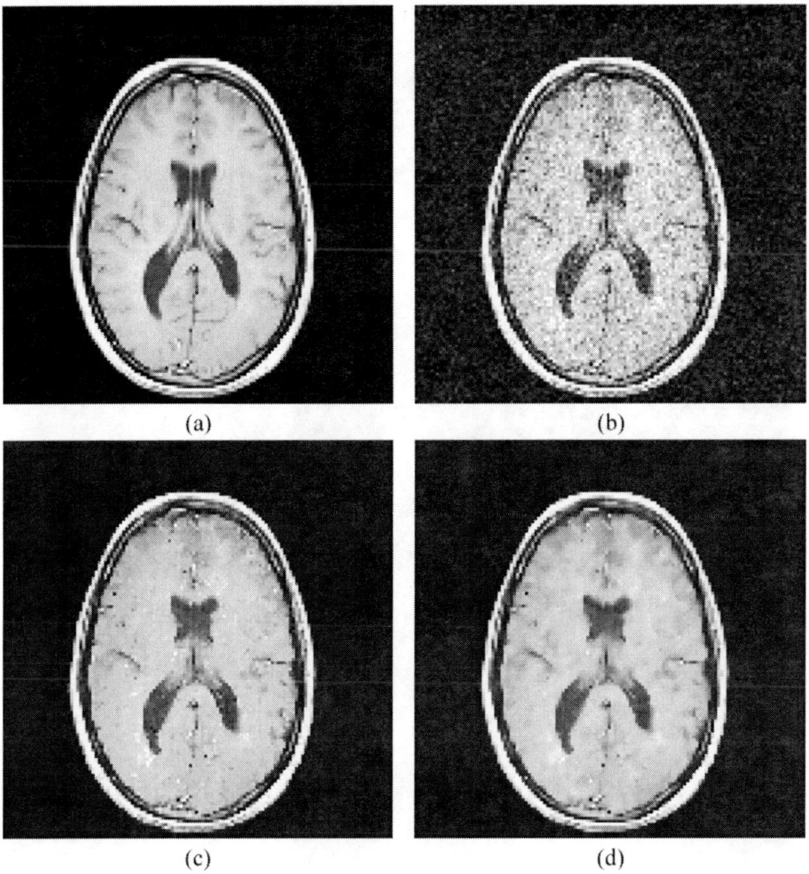

Figure 2. Comparing the results of LPA-ICI-GA *versus* LPA-ICI algorithms for brain MRI slice restored images after adding Rician noise with s = 0.1 (noise variance). (**a**) is the Brain MRI original image, (**b**) Corrupted image with Rician noise with noise variance (s = 0.1), (**c**) is the de-noised image based LPA-ICI algorithm, and (**d**) is the de-noised image based LPA-ICI supported by the genetic algorithm (proposed system LPA-ICI-GA).

Another comparison set of the brain MRI slice images is shown in Figure 3 for the same original image shown in Figure 2a after adding Rician noise with different variance levels. This figure shows a different comparison using the proposed (LPA-ICI-GA) and the LPA-ICI compared to the original brain MRI image after adding the Ricicn noise. The obtained results demonstrate the superiority of the proposed method over the LPA-ICI algorithm even with high noise variance values. At high noise variance, the images become not clear using LPA-ICI.

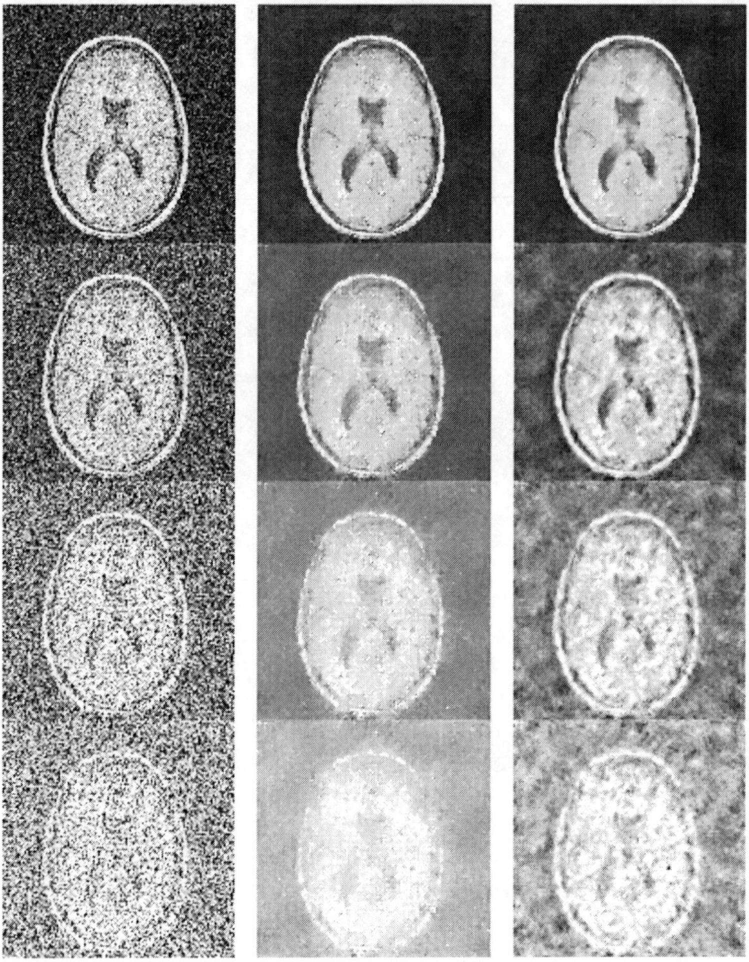

Figure 3. Comparing the results of LPA-ICI-GA *versus* LPA-ICI algorithms for brain MRI slice restored images after: (**Left panel**) adding Rician noise to the original image in Figure 2a at variance levels (s = 0.2, 0.3, 0.4, and 0.5, respectively); (**Middle panel**) restored images using LPA-ICI, (**Right panel**) restored images using LPA-ICI-GA.

5.2. Proposed System Performance Analysis

Assume here that Z(i,j) is the original image of siz M1×M2, while X(i,j) is the image restored using LPA-ICI with parameters given by the GA. Additionally, the fitness function (FF) is given by PSNR+100*MSSIM. To evaluate the system performance, various performance metrics are measured for both the proposed method (LPA-ICI-GA) and the LPA-ICI method. These metrics are SNR (Signal-to-Noise Ratio), PSNR (Peak SNR), ISNR (Improvement in SNR), MSE (Mean Squared Error), RMSE (Root of MSE), MAE (Mean Absolute Error), MAX (Maximum Absolute Difference). These metrics are defined as:

- Signal-to-noise ratio (SNR): is defined as the ratio of the power of the original image values and the power of noise values. It is given by:

$$SNR = 10 \times \log_{10} \times \left. \frac{\sum\sum X(i,j)^2}{\sum\sum (Z(i,j) - X(i,j))^2} \right| \quad (11)$$

- The improvement in signal to noise ratio (ISNR) between the original and restored images is given by:

$$ISNR = 10 \times \log_{10} \times \left. \frac{\sum\sum (X(i,j) - Y(i,j))^2}{\sum\sum (Z(i,j) - X(i,j))^2} \right| \quad (12)$$

- The peak signal to noise ratio (PSNR) between the original and restored images is given by [68]:

$$PSNR = 20 \times \log_{10} \times \left. \frac{255 \times M \times N}{\sqrt{\sum\sum (Z(i,j) - X(i,j))^2}} \right| \quad (13)$$

- The mean squared error (MSE): measures the average of the squares of the "errors", known as the difference between the original image and the restored image. It is given by,

$$MSE = \frac{\sum\sum (Z(i,j) - X(i,j))^2}{M \times N} \tag{14}$$

- The mean absolute error (MAE): is the average of the absolute errors between the original image and the restored image. It is given by:

$$MAE = \frac{1}{M \times N} \sum\sum |Z(i,j) - X(i,j)| \tag{15}$$

- The maximum absolute difference (MAD) is given by:

$$MAD = \max(|Z(i,j) - X(i,j)|) \tag{16}$$

- The SSIM index measure between two windows Z and X of the original and restored images is given by:

$$SSIM(x,z) = \frac{(2\mu_x\mu_z + C_1)(2\sigma_{xz} + C_2)}{(\mu_x^2 + \mu_z^2 + C_1)(\sigma_x^2 + \sigma_z^2 + C_2)} \tag{17}$$

where, μx is the average of *X*, μz is the average of *Z*, σx2 is the variance of *X*, σz2 is the variance of *Z*, σxz is the covariance of *X* and *Z*, C1=(K1L)2 and C2=(K2L)2 are two variables used to stabilize the division, L = 255 is the dynamic range of pixel-values, and K1 = 0.01 and K2 = 0.03 by default. The SSIM is calculated on various windows of the images.

- The mean structural similarity (MSSIM) index is given by:

$$MSSIM(X,Z) = {}^{1}\!/_{P} \sum_{J} SSIM(x_j, z_j) \tag{18}$$

where, xj and zj are the image contents at the jth local window, and P is the number of local windows in the image.

Along with these parameters with different Rician noise variance (s), the following results calculated.

Table 1 illustrates theses performance metrics when using the LPA-ICI without optimization; where the values used for the ICI parameters are fixed and are selected as to have the values: Sharpness parameter = −1, Γ = 1.05, Directional resolution = 8 and Fusing = 1 [65], at different values of the noise variance "s". Where these parameters are set to be fixed when using the LPA based ICI rule, then adjust the window size according to theses fixed value parameters which affect its performance. Where, the LPA-ICI multi-directional kernels give additional enhancement, providing an efficient tool especially for image de-noising, differentiation and inverse-imaging problems.

The same metrics are measured for the same original test image used in Figure 2 and Figure 3 at different Rician noise variance "s", when using the proposed LPA-ICI-GA that provides optimal parameter values that used in the ICI rule to determine the optimal window width. The optimal parameters values and the measured metrics are shown in Table 2 and Table 3.

Since, the proposed algorithm uses the genetic algorithm to optimize the ICI parameters, therefore a range for each parameter to be used. A Matlab Genetic Algorithm toolbox is used in the GA process. For these parameters tuning, selected random values for all the parameters in the integer numbers form are used as a combination set. It is noted that the authors in [65] test their algorithm for Γ range from 1.5 to 4.0. As, Γ is the threshold for the confidence interval used in the ICI rule to obtain the adaptive window size. Jointly with the fusing has only two possible values, where fusing = 1 for classical estimation, and equal 2 for piecewise estimation. In addition, the LPA-ICI uses the directional parameter to choose the number of directions in which the kernels will be applied. Actually, its range can be chosen to be integer from 1 to 8, but the most common values for the directional resolution are 4 and 8 neighborhood for a pixel. Generally, regarding the sharpening, −1 refers to "no sharpening", and if greater than 0 refers to the "sharpening", it multiplies the value with the kernel matrix to improve sharpness. Therefore, the parameter selection range will be:

Table 1. The performance metrics values for the LPA-ICI without GA.

S	ISNR	SNR	PSNR	MSE	RMSE	MAE	MAX	MSSIM
0.1	1.55	12.2267	19.5451	722.0499	26.871	23.7222	95.379	0.44208
0.2	1.6386	6.31	13.6284	2819.9239	53.103	46.5034	190.9031	0.31629
0.3	1.6375	2.8197	10.1381	6299.0275	79.3664	69.6002	245.642	0.22051
0.4	1.612	0.31846	7.6369	11204.5665	105.8516	93.8354	367.7877	0.15175
0.5	1.5691	-1.6638	5.6546	17685.5917	132.9872	119.8152	463.6092	0.10741
0.6	1.5325	-3.3066	4.0118	25816.7765	160.676	147.2261	559.6435	0.084966
0.7	1.4968	-4.718	2.6004	35730.901	89.0262	175.6339	655.8092	0.070399

- Sharpness parameter range = [−1, from 0 till 10];
- Γ range = [from 0 till 5];
- Directional resolution range = [[4 or 8];
- Fusing range = [[1 or 2], where Fusing = 1 for classical estimation, and equal 2 for piecewise estimation.

As clear from Table 2 and Table 3 that:

Step 1. According to the noise variance level, the ICI parameter optimal values are changeable. This proves the importance of the GA to obtain the optimized values for these parameters. Also, these parameter values are changeable with any change in the image concerned.

Step 2. The ICI tuned parameters (optimized), are different than the fixed values used in LPA-ICI. Also, as the noise variance increase and equal 0.6 both fusing and the directional resolution switches their values, to be able to overcome the high noise level.

Step 3. Comparing the performance metrics values in Table 1 and Table 3, it is clear that, the proposed method is outperforming the LPA-ICI in terms of all the metric values.

Table 2. The optimal ICI parameter values when using LPA-ICI-GA for MRI image de-noising.

S	SharpParam	Γ	Directional Resolution	Fusing
0.1	−0.4495	0.4984	8	2
0.2	−0.7464	0.7165	8	2
0.3	0.0126	0.6696	8	2
0.4	−0.1287	0.5449	8	2
0.5	−0.0556	0.5540	8	2
0.6	−0.2439	0.5234	8	2
0.7	−0.8946	2.3293	4	1

Table 3. The performance metrics values when using LPA-ICI-GA for MRI image de-noising.

S	ISNR	SNR	PSNR	MSE	RMSE	MAE	MAX	MSSIM	Fitness Function = PSNR+100*MSSIM
0.1	1.5982	12.275	19.5934	714.0724	26.7221	23.4761	95.379	0.44871	64.4645
0.2	1.6994	6.3708	13.6892	2780.722	52.7325	45.9734	190.9031	0.32608	46.2971
0.3	1.8495	3.0317	10.3501	5998.932	77.4528	68.2574	211.127	0.24197	34.5472
0.4	1.822	0.52845	7.8469	10,675.67	103.3231	91.4624	238.5945	0.18155	26.0015
0.5	1.7566	−1.4763	5.8421	1,6938.2	130.1469	116.3874	297.4721	0.13606	19.4483
0.6	1.6849	−3.1542	4.1642	2,4926.53	157.8814	143.7663	312.2485	0.10795	14.9587
0.7	1.4848	−4.73	2.5884	3,5829.55	189.2869	176.1798	363.922	0.085832	11.1716

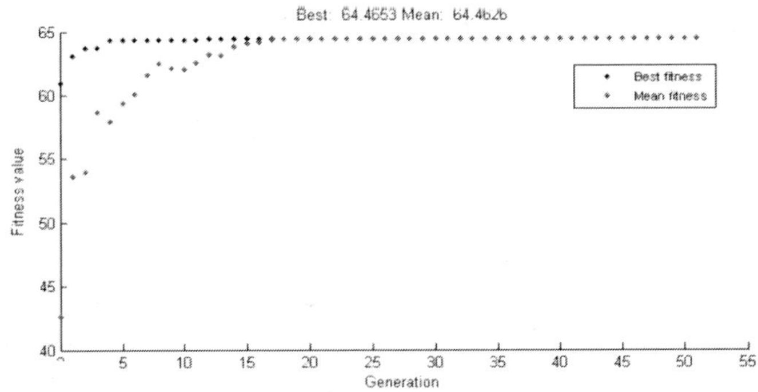

(a)

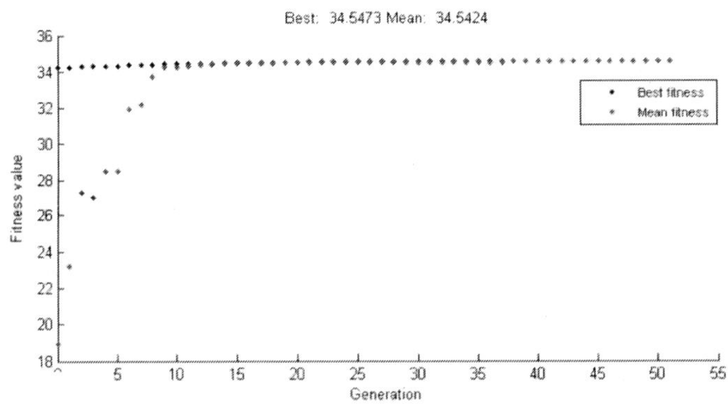

(b)

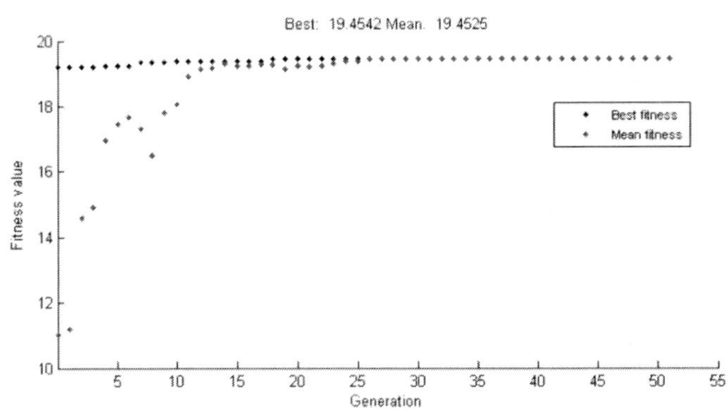

(c)

Figure 4. The fitness-convergence graph at different noise variance values (s= **(a)** 0.1, **(b)** 0.3 and **(c)** 0.5).

5.3. Proposed System Convergence Graphs

Figure 4 is related to Table 2 as it shows the obtained convergence graph where the graph reports the number of iterations (generations) *versus* the fitness function when using the GA. It is obvious from Figure 4 that, as the noise variance level increase, the fitness function value decreases. Also, the range of number of generations (iterations) required for convergence is about 25 iterations maximum and it varies according to the noise variance value.

In literature [69] by Iftikhar *et al.*, similar work has been done in recent past. The authors employed Elite GA (EGA) to find the optimal values that the Non-local means (NLM) filter depends on the patch size, search window size and the smoothing parameters. It was reported that for given images, the GA converges well before 50 iterations using the PSNR as fitness value. While, using the EGA accelerate the search process with better convergence rate. The authors tested images of different noise level variance (10, 15 and 20). For synthetic images, it was stated that at variance = 10, the GA converged at 32 iterations while the EGA converged at 26 iterations. While, with variance = 15 or 20, both the GA and EGA converged at 21 iterations. With respect to the PSNR the GA and EGA recorded the same 35.87, 32.80, and 31.67 for variance values of 10, 15, and 20; respectively. Moreover, with respect to the RMSE both algorithms reported the same values which were 3.77, 5.28, and 6.29 for variance values of 10, 15, and 20; respectively.

Where as in our work using LPA-ICI-GA algorithm, the variance level was in the range from 0.1 to 0.7, where the LPA-ICI-GA algorithm converged after about 25 iterations. As established by Table 3, the PSNR equal 19.5934, 13.6892, and 10.3501 with s = 0.1, 0.2, and 0.3; respectively. Furthermore, the RMSE had values of 26.7221, 52.7325, and 77.4528 with s = 0.1, 0.2, and 0.3; respectively.

Comparing the results obtained by our proposed algorithm to that obtained by Iftikhar *et al.*[69]. It is clear that, even if the dataset, variance level, and fitness function are different in Iftikhar *et al.* [69] from those used in the current proposed technique, it can be concluded that our obtained results are very much encouraging and promising.

Moreover, a lot of work has been done in recent time in the area of MRI de-noising. The following table (Table 4) shows some of these related methods indicating the de-noising techniques used and compared to the proposed method.

Table 4. Comparison table of related work for MRI images de-noising and the current study.

S. No.	Authors	Year	De-Noising Technique	Comments/Notes
1	Bao *et al.* [15]	2003	Adaptive wavelet thresholding	To, this method multiplied the adjacent wavelet sub bands, then applied threshold to the multi-scale products to differentiate edge structures in an improved approach from noise and amplify the significant features. Results of MRI images proved that this method achieved high mean-to-standard-deviation ratio (MSR) and contrast to noise ratio (CNR).
2	Wang *et al.* [18]	2006	Wavelet and the total variation minimization methods	An effective automatic stopping time criterion extensively.
3	Manjon *et al.* [19]	*2009*	Novel filter averaging similar patches around the image using a robust multicomponent similarity measure filter, then local Principal Component Analysis decomposition as a post-processing for more noise removal.	This method showed a consistent improvement in the results when the number of images increases in contrast with the erratic behavior of the basic version.
4	Jaya *et al.* [23]	2009	Median filter, the weighted median and the adaptive filter.	The results proved that weighted median filter was outperforming the other filters with peak signal to noise ratio PSNR = 0.924.
5	Rajeesh *et al.* [20]	2010	Curvelet shrinkages were superior than using the wavelet method.	The reconstructed MRI data have high Signal to Noise Ratio (SNR) compared to the curvelet and wavelet domain denoising approaches.
6	Erturk *et al.* [22]	2013	Spectral subtraction method.	The results showed enhanced signal to noise ratio (SNR) up to 40% in the MRI reconstructed signal
7	Proposed system	2015	Local polynomial approximation with intersection confidence interval based genetic algorithm filter. (LPA-ICI-GA)	The results proved that using GA optimization to support the LPA-ICI outperform the classic LPA-ICI. For variance of 0.1, the PSNR using the proposed method is 19.5934 dB, while that obtained by the LPA-ICI was 19.5451. Comparing with respect to the MSE at the same variance value, it was obtained that the LPA-ICI-GA compared to the LPA-ICI achieved 714.0724 and 722.0499; respectively. That established that the proposed method gained les MSE compared to the classical LPA-ICI approach.

It is clear from Table 4 that, most of the work in the literature were used conventional de-noising techniques, while the selection of any filter parameters are considered an optimization problem. In this work the main motivation is to use optimization algorithm to adjust the parameters of the classic LPA-ICI filter, *i.e.*, the optimal window size selection, for MRI images de-noising. Thus, the GA optimization approach is used in order to reduce the Rician noise by optimizing the mean square error taken as a cost function. The main contribution of this work can be pointed as: (I) used GA for optimal window size selection for the classic LPA-ICI, (II) provided the optimal window size with changeable image under concern, (III) compared the performance metrics' values of the proposed system to those for the classic LPA-ICI system, (IV) proved that the proposed method is outperforming the LPA-ICI in terms of all the metric values, (V) tested the proposed novel idea with different Rician noise variance levels, (VI) used the brain MRI images corrupted with Rician noise to test the proposed system, (VII) achieved about 25 iteration maximum for convergence and it varies according to the noise variance value, (IX) established the superior efficiency of the proposed LPA-ICI-GA over the classic LPA-ICI for MRI images de-noising, (X) for variance of 0.1, the PSNR using the proposed method is 19.5934 dB, while that obtained by the LPA-ICI was 19.5451. Comparing with respect to the MSE at the same variance value, it was obtained that the LPA-ICI-GA compared to the LPA-ICI achieved 714.0724 and 722.0499; respectively. That established that the proposed method gained less MSE compared to the classical LPA-ICI approach, and (XI) endowed with a high level of noise reduction, it can give good performance till noise variance level of 0.5.

Due to the efficient performance of the proposed system, it can be generalized to be used with any type of medical/ non-medical images for de-noising and restore the images.

6. CONCLUSIONS AND FUTURE WORK

Noise is a significant factor that can maneuver medical image quality. Medical images are vulnerable to noise affect in the process of transmission and processing which cause the loss of image contrast, blur, *etc.* Physicians will not be able to discriminate the interested lesions, thus influence the diagnosis, when the diseased tissue and normal tissue attenuation coefficient is very small. So, filtering out the signal noise, getting a restored noise free image, become the physician's main concern. In recent year, Medical image de-noising is an evolving research area that has obtained vast consideration among the researchers. Significant researches and techniques that exist for medical image de-noising are pursued.

Due to the nonparametric characteristic of the local polynomial approximation filter and the frequency distribution characteristics of noise, LPA-ICI had a superior benefit in image de-noising. In this work to improve this method an optimization algorithm is to be used to optimize the ICI parameters that used later to determine the sliding window size. Since, the Genetic Algorithm uses a bulky number of solutions, rather than a single solution for searching. So, GA is a robust procedure that improves the chance of realizing the global optimum and nearly unbiased optimization techniques for sampling a large solution space. Thus, the GA is used to support the LPA-ICI de-noising filter.

This proposed method is tested using brain MRI images corrupted with Rician noise. Where, MRI has developed appreciably over the last few decades. It is commonly well recognized to model noise on magnitude MR images as white and Rician distributed.

An innovative approach to de-noise Rician noise has been conducted in this work which use of the Genetic Algorithm to optimize the ICI parameters using a fitness function. The GA proves its efficiency in this proposed method as it achieves better results, fast processing times and easy implementation with satisfactory performance compared to using LPA-ICI without optimization. In addition to, the proposed method endowed with a high level of noise reduction, it can give good performance till noise variance level of 0.5.Current work reports better PSNR, ISNR, SNR, MSE, MAE, and MAX values.

For further work, it is suggested to change the parameters range used study its effect on the proposed method. For example, it is possible to check if the Directional resolution range has other values rather than 4 and 8 only.

Furthermore, it is suggested to compare the proposed method performance with that obtained using wavelet and wavelet based GA and compares the computed time efficiency. In addition, the proposed algorithm can be tested with any other medical image modalities. Besides, it is recommended to use different optimization algorithms either heuristic or meta-heuristic and compare their performance when used with the LPA-ICI or RICI de-noising filters.

ACKNOWLEDGMENTS

The authors would like to thank the Editor and anonymous reviewers for their valuable comments and suggestions, which were helpful in improving the paper.

AUTHOR CONTRIBUTIONS

Nilanjan Dey and Amira Ashour conceived and designed the experiments, and then Samsad Beagum performed the experiment under the guidance of Nilanjan Dey and Amira Ashour. Nilanjan Dey and Amira Ashour analyzed the results/data with the assistant of Mitko Gospodinov and Evgeniya Peneva Gospodinova who also contributed to provide the dataset and all necessary tools. Finally, Nilanjan Dey and Amira Ashour and João Manuel Tavares draft the manuscript, and afterward Dimitra Sifaki Pistola revised the whole manuscript grammatically.

CONFLICTS OF INTEREST

The authors declare that no conflict of interest.

REFERENCES

1. Dey, N.; Das, P.; Roy, A.; Das, A.; Chaudhuri, S. Detection and measurement of Arc of Lumen calcification from intravascular ultrasound using Harris corner detection. In Proceedings of National Conference on Computing and Communication Systems (NCCCS), Durgapur, India, 21–22 November 2012.
2. Dey, N.; Roy, A.; Pal, M.; Das, A. FCM based blood vessel segmentation method for retinal images. *Int. J. Comput. Sci. Netw.* **2012**, *1*.
3. Araki, T.; Ikeda, N.; Molinari, F.; Dey, N.; Acharjee, S.; Saba, L.; Nicolaides, A.; Suri, J. Effect of geometric-based coronary calcium volume as a feature along with its shape-based attributes for cardiological risk prediction from low contrast IVUS. *J. Med. Imaging Health Inform.* **2014**, *4*, 255–261.
4. Dey, N.; Dey, M.; Biswas, D.; Das, P.; Das, A.; Chaudhuri, S. Tamper detection of electrocardiographic signal using watermarked bio-hash code in wireless cardiology. *Int. J. Signal Imaging Syst. Eng.* **2015**, *8*, 46–58.
5. Bilcu, R.; Vehvilainen, M. A novel decomposition scheme for image de-noising. In Proceedings of the IEEE International Conference on Acoustics, Speech and Signal Processing, Honolulu, HI, USA, 15–20 April 2007; pp. 577–580.
6. Alpuente, L.; López, A.; Tur, R. Glioblastoma: Changing expectations? *Clin. Trans. Oncol.***2011**, *13*, 240–248.

7. Sijbers, J.; Dekker, A.; Audekerke, J.; Verhoye, M.; Dyck, D. Estimation of the noise in magnitude MR images. *Magn. Reson. Imaging* **1998**, *16*, 87–90.
8. Rice, S. Mathematical analysis of random noise. *Bell Syst. Tech. J.* **1944**, *23*, 282–332.
9. Ali, S.; Vathsal, S.; Lalkishore, K. A GA-based window selection methodology to enhance window-based multi-wavelet transformation and thresholding aided CT image denoising technique. *Int. J. Comput. Sci. Inf. Secur.* **2010**, *7*.
10. Ali, S.A.; Vathsal, S.; Kishore, K.L. CT image denoising technique using GA aided window based multiwavelet transformation and thresholding with the incorporation of an effective quality enhancement method. *Int. J. Digit. Content Technol. Appl.* **2010**, *4*, 75–87.
11. Dey, N.; Das, A.; Chaudhuri, S. Wavelet based normal and abnormal heart sound identification using spectrogram analysis. *Int. J. Comput. Sci. Eng. Technol.* **2012**, *3*, 186–192.
12. Dey, N.; Roy, A.; Das, A.; Chaudhuri, S. Stationary wavelet transformation based self-recovery of blind-watermark from electrocardiogram signal in the wireless telecardiology. In Proceedings of the International Workshop on Intelligence and Security Informatics for International Security (IIS'12), Trivandrum, India, 11–12 October 2012.
13. Healy, D.; Weaver, J. Two applications of wavelet transforms in magnetic resonance imaging. *IEEE Trans. Inf. Theory* **1992**, *38*, 840–860.
14. Nowak, R. Wavelet-based Rician noise removal for magnetic resonance imaging. *IEEE Trans. Image Process.* **1999**, *8*, 1408–1419.
15. Bao, P.; Zhang, L. Noise reduction for magnetic resonance images via adaptive multiscale products thresholding. *IEEE Trans. Med. Imaging* **2003**, *22*, 1089–1099.
16. Jiang, L.; Yang, W. Adaptive magnetic resonance image denoising using mixture model and wavelet shrinkage. In Proceeding of the VIIth Digital Image Computing: Techniques and Applications, Sydney, Australia, 10–12 December 2003; pp. 831–838.
17. Kadah, Y. Adaptive denoising of event-related functional magnetic resonance imaging data using spectral subtraction. *IEEE Trans. Biomed. Eng.* **2004**, *51*, 1944–1953.
18. Wang, Y.; Zhou, H. Total variation wavelet-based medical image denoising. *Int. J. Biomed. Imaging* **2006**, *2006*. Article ID 89095.
19. Manjon, J.; Thacke, N.; Lull, J.; Mar, G.; Bonmat, L.; Robles, M. Research article multicomponent MR image denoising. *Int. J. Biomed. Imaging* **2009**, *2009*, 1–10.

20. Rajeesh, J.; Moni, R.; Palanikumar, S.; Gopalakrishnan, T. Noise reduction in magnetic resonance images using wave atom shrinkage. *Int. J. Image Process.* **2010**, *4*, 131–141.
21. Balafar, M. Review of noise reducing algorithms for brain MRI images. *Int. J. Tech. Phys. Probl. Eng.* **2012**, *4*, 54–59.
22. Erturk, M.; Bottomley, P.; El-Sharkawy, A. Denoising MRI using spectral subtraction.*IEEE Trans. Biomed. Eng.* **2013**, *60*, 1556–1562.
23. Jaya, J.; Thanushkodi, K.; Karnan, M. Tracking algorithm for de-noising of MR brain images. *Int. J. Comput. Sci. Netw. Secur.* **2009**, *9*, 262–267.
24. Iftikhar, M.A.; Jalil, A.; Rathore, S.; Ali, A.; Hussain, M. Brain MRI denoizing and segmentation based on improved adaptive nonlocal means. *Int. J. Imaging Syst. Technol.***2013**, *23*, 235–248.
25. Jalil, A.; Rathore, S.; Hussain, M. Robust brain MRI denoising and segmentation using enhanced non-local means algorithm. *Int. J. Imaging Syst. Technol.* **2014**, *24*, 52–66.
26. Iftikhar, M.A.; Jalil, A.; Rathore, S.; Ali, A.; Hussain, M. An extended non-local means algorithm: Application to brain MRI. *Int. J. Imaging Syst. Technol.* **2014**, *24*, 293–305.
27. Klepaczko, A. *AGPU Accelerated Local Polynomial Approximation Algorithm for Efficient Denoising of MR Images*; Burduk, R., Jackowski, K., Kurzynski, M., Wozniak, M., Zolnierek, A., Eds.; Springer International Publishing: Cham, Switzerland, 2013.
28. Coupé, P.; Manjón, J.; Gedamu, E.; Arnold, D.; Robles, M.; Collins, D.L. Robust Rician noise estimation for MR images. *Med. Image Anal.* **2010**, *14*, 483–493.
29. Tan, X.; Sun, C.; Pham, T.D. Multipoint filtering with local polynomial approximation and range guidance. In Proceeding of the 2014 IEEE Conference in Computer Vision and Pattern Recognition (CVPR), Columbus, OH, USA, 23–28 June 2014; pp. 2941–2948.
30. Hu, Y.; Jiang, X.; Xin, F.; Zhang, T.; Yuan, J.; Zhai, L.; Guo, C. An algorithm on processing medical image based on rough-set and genetic algorithm. In Proceedings of the International Conference on Information Technology and Applications in Biomedicine, Shenzhen, China, 30–31 May 2008.
31. Samanta, S.; Dey, N.; Das, P.; Acharjee, S.; Chaudhuri, S. Multilevel threshold based gray scale image segmentation using Cuckoo search. In Proceedings of the International Conference on Emerging Trends in Electrical, Communication and Information Technologies (ICECIT), Anantapur, India, 12–23 December 2012.
32. Chakraborty, S.; Pal, A.; Dey, N.; Das, D.; Acharjee, S. Foliage area computation using Monarch butterfly algorithm. In Proceedings of the 2014 1st International Conference on Non Conventional Energy, Kalyani, India, 16–17 January 2014.

33. Jayashri, P.; Gandhimathi, D. Efficient tumor segmentation in medical images using artificial bee colony optimization algorithm and fuzzy c-means clustering. *Int. J. Adv. Res. Comput. Sci. Manag. Stud.* **2014**, *2*, 98–101.
34. Acharjee, S.; Dey, N.; Samanta, S.; Das, D.; Roy, R.; Chakraborty, S.; Chaudhuri, S. ECG signal compression using ant weight lifting algorithm for tele-monitoring. *J. Med. Imaging Health Inform.* **2014**, *5*, 1580–1587.
35. Day, N.; Samanta, S.; Chakraborty, S.; Das, A.; Chaudhuri, S.; Suri, J. Firefly algorithm for optimization of scaling factors during embedding of manifold medical information: An application in ophthalmology imaging. *J. Med. Imaging Health Inform.* **2014**, *4*, 384–394.
36. Misra, D.; Sarker, S.; Dhabal, S.; Ganguly, A. Effect of using genetic algorithm to denoise MRI images corrupted with Rician Noise. In Proceedings of the 2013 IEEE International Conference on Emerging Trends in Computing, Communication and Nanotechnology, Tirunelveli, India, 25–26 March 2013; pp. 146–151.
37. Liu, Y.; Ma, Y.; Liu, F.; Zhang, X.; Yang, Y. The research based on the genetic algorithm of wavelet image denoising threshold of medicine. *J. Chem. Pharm. Res.* **2014**, *6*, 2458–2462.
38. Tsang, P.; Au, A. A genetic algorithm for projective invariant object recognition. In Proceedings of the 1996 IEEE TENCON: Digital Signal Processing Applications, Perth, WA, USA, 26–29 November 1996; pp. 58–63.
39. Yang, J.; Fan, J.; Ai, D.; Zhou, S.; Tang, S.; Wang, Y. Brain MR image denoising for Rician noise using pre-smooth non-local means filter. *Biomed. Eng. Online* **2015**, *14*, 1–20.
40. Gudbjartsson, H.; Patz, S. The Rician distribution of noisy MRI data. *Magn. Reson. Med.***1995**, *34*, 910–914.
41. Jagadeesan, H.; Sasirekha, N. Robust Rician noise estimation and filtering for magnetic resonance imaging. *Int. J. Sci. Eng. Res.* **2014**, *5*, 620–624.
42. Selvathi, D.; Selvi, S.; Malar, C. The SURE-LET approach for MR brain image denoising using different shrinkage rules. *Int. J. Healthc. Inf. Syst. Inform.* **2010**, *5*, 73–81.
43. Edelstein, W.; Glover, G.; Hardy, C.; Redington, R. The intrinistic signal-to-noise ratio in MR imaging. *Magn. Reson. Med.* **1986**, *3*, 604–618.
44. Macovski, A. Noise in MRI. *Magn. Reson. Med.* **1996**, *36*, 494–497.
45. Basu, S.; Fletcher, T.; Whitaker, R. Rician noise removal in diffusion tensor MRI. In*Medical Image Computing and Computer-Assisted Intervention—MICCAI 2006*; Larsen, R., Nielsen, M., Sporring, J., Eds.; Springer: Berlin, Germany, 2006; pp. 117–125.
46. Papoulis, A. *Probability, Random Variables, and Stochastic Processes*; McGraw-Hill: New York, NY, USA, 1984.

47. Nobi, M.; Yousuf, M. A new method to remove noise in magnetic resonance and ultrasound images. *J. Sci. Res.* **2011**, *3*, 81–89.
48. Lee, J.S. Digital image enhancement and noise filtering by use of local statistics. *IEEE Trans. Pattern Anal. Mach. Intell.* **1980**, *2*, 165–168.
49. Russ, J. *The Image Processing Handbook*, 6th ed.; CRC Press: Baca Raton, FL, USA, 1999.
50. Lee, Y.; Rhee, S. Wavelet-based image denoising with optimal filter. *Int. J. Inf. Process. Syst.* **2005**, *1*, 32–35.
51. Gupta, V.; Mahle, R.; Shriwas, R.S. Image denoising using wavelet transform method. In Proceedings of the 2013 Tenth International Conference on Wireless and Optical Communications Networks (WOCN), Bhopal, India, 26–28 July 2013.
52. Weaver, D.; Xu, Y.; Driscoll, J. Filtering MR images in the wavelet transform domain.*Magn. Reson. Med.* **1991**, *21*, 288–295.
53. Hilton, M.; Ogden, T.; Hattery, D.; Eden, G.; Jawerth, B. Wavelet de-noising of functional MRI data. In *Wavelets in Biology and Medicine*; Aldroubi, A., Unser, M., Eds.; CRC Press: Baca Raton, FL, USA, 1996; pp. 93–112.
54. Piˇzurica, A.; Philips, W.; Lemahieu, I.; Acheroy, M. A versatile wavelet domain noise filtration technique for medical imaging. *IEEE Trans. Med. Imaging* **2003**, *22*, 323–331.
55. Katkovnik, V. A new method for varying adaptive bandwidth selection. *IEEE Trans. Signal Process.* **1999**, *47*, 2567–2571.
56. Ashour, A.; Elkamchouchi, H. Enhancement of moving targets tracking performance using the ICI rule. *Alex. Eng. J.* **2007**, *46*, 673–682.
57. Perona, P.; Malik, J. Scale-space and edge detection using anisotropic diffusion. *IEEE Trans. PAMI* **1990**, *12*, 629–639.
58. Li, S.; Huang, D. Image denoising using non-negative sparse coding shrinkage algorithm. In Proceedings of the IEEE Computer Society Conference on Computer Vision and Pattern Recognition, San Diego, CA, USA, 20–25 June 2005; pp. 1017–1022.
59. Crouse, M.; Nowak, R.; Baraniuk, R. Wavelet-based statistical signal processing using hidden markov models. *IEEE Trans. Signal Process.* **1998**, *46*, 886–902.
60. Jangra, S.; Yadav, S. A review of Rician noise reduction in MRI images using wave atom transform. *Int. J. Comput. Sci. Mob. Comput.* **2014**, *3*, 454–457.
61. Katkovnik, V.; Egiazarian, K.; Astola, J. *Local Approximation Techniques in Signal and Image Processing*; SPIE Press: Bellingham, WA, USA, 2006.
62. Katkovnik, V.; Egiazarian, K.; Astola, J. A spatially adaptive nonparametric regression image deblurring. *IEEE Trans. Image Process.* **2005**, *14*, 1469–1478.

63. Fan, J.; Gijbels, I. *Local Polynomial Modelling and Its Application*; Chapman and Hall: London, UK, 1996.
64. Katkovnik, V. *The Method of Local Approximation*; Nauka: Moscow, Russia, 1985.
65. Katkovnik, V.; Egiazarian, K.; Astola, J. Adaptive window size image de-noising based on Intersection of Confidence Intervals (ICI) rule. *J. Math. Imaging Vis.* **2002**, *16*, 223–235.
66. Malhotra, R.; Singh, N.; Singh, Y. Genetic algorithms: Concepts, design for optimization of process controllers. *Comput. Inf. Sci.* **2011**, *4*.
67. Exploring Slices from a 3-Dimensional MRI Data Set. Available online: http://in.mathworks.com/help/images/examples/exploring-slices-from-a-3-dimensional-mri-data-set.html(accessed on 19 August 2015).
68. Chen, T.; Wu, H. Space variant median filters for the restoration of impulse noise corrupted images. *IEEE Trans. Circuits Syst. II* **2011**, *48*, 784–789.
69. Iftikhar, A.; Rathore, S.; Jalil, A. Parameter optimization for non-local de-noising using Elite GA. In Proceedings of the 15th International Multitopic Conference (INMIC), Islamabad, Pakistan, 13–15 December 2012; pp. 194–199.

5
CHAPTER

FUSION OF VISIBLE AND THERMAL DESCRIPTORS USING GENETIC ALGORITHMS FOR FACE RECOGNITION SYSTEMS

Gabriel Hermosilla [1,*], Francisco Gallardo [1], Gonzalo Farias [1] and Cesar San Martin [2]

[1] *Escuela de Ingeniería Eléctrica, Pontificia Universidad Católica de Valparaíso, Valparaíso 2362804, Chile*

[2] *Department of Electrical Engineering, University of La Frontera, Temuco 4811230, Chile*

ABSTRACT

The aim of this article is to present a new face recognition system based on the fusion of visible and thermal features obtained from the most current local matching descriptors by maximizing face recognition rates through the use of genetic algorithms. The article considers a comparison of the performance of the proposed fusion methodology against five current face recognition methods and classic fusion techniques used commonly in the literature. These were selected by considering their performance in face recognition. The five local matching methods and the proposed fusion methodology are evaluated using the standard visible/thermal database, the Equinox database, along with a new database, the PUCV-VTF, designed for visible-thermal studies in face recognition and described for the first time in this work. The latter is created considering visible and thermal image sensors with different real-world conditions, such as variations in illumination, facial expression, pose, occlusion, *etc.* The main conclusions of this article are that two variants of the proposed fusion methodology surpass current face recognition methods and the classic fusion techniques reported in the literature, attaining recognition rates of over 97% and 99% for the Equinox and PUCV-VTF databases, respectively. The fusion methodology is very robust to illumination and expression changes, as it combines thermal and visible information efficiently by using genetic algorithms, thus allowing it to choose optimal face areas where one spectrum is more representative than the other.

Keywords: face recognition; fusion descriptors; genetic algorithms; visible and infrared spectrum

1. INTRODUCTION

Face recognition in the visible spectrum is a well-known field of research with many years of study using low-cost cameras. Many publications are available, containing smart and sophisticated algorithms. This field has seen steady growth due to its security applications, as well as other types of applications, such as access permission and even identity control. However, one of the main problems that researchers must overcome is the diminished recognition ability of algorithms as a result of variations in the intensity of illumination in the image. In a real application, where intensity of illumination can vary due to factors, such as the time or weather, face recognition with visible imagery may no longer be robust (*i.e.*, low recognition rates).

A solution to the problem caused by lighting in visible imagery is to use images acquired in the infrared (IR) spectrum (specifically, the thermal spectrum, 8–12 μm), because the images obtained in this spectral range remain invariant to lighting changes that may occur in the visible spectrum [1]. This is possible because thermal cameras capture the infrared radiation emitted by the face instead of the reflected light, following Stefan Boltzmann's law. The estimation of the temperature of a body is a process called thermography [2], in which it is not necessary to make contact with the surface of the body.

A disadvantage of the IR spectrum is that the energy received by the IR camera not only depends of the measured body, but also on the sum of the energy components of the different elements of the scene captured by the camera. The elements of the scene that are captured are: the atmosphere, the background and the object to be measured. The variation of these components could affect the temperature estimation of the measured body, faces in this case, which is a problem in face recognition. The thermal face images also have undesirable variations due to changes in ambient temperature, alterations of the metabolic processes of the subjects, camera susceptibility to extrinsic factors and variable sensor responses over time when the camera is working for long periods of time [3,4].

Based on the advantages and disadvantages of both spectra, it is possible to create a face recognition system based on the fusion of the visible and thermal domain, achieving a high recognition rate. The aim of this article is described in the following three points: (1) the article proposes a new fusion system that combines visible and thermal features obtained from the most current local matching descriptors by maximizing face recognition rates through the use of genetic algorithms (GA, [5,6]); (2) the article compares and studies current face recognition methods and classic fusion approaches against the proposed fusion technique; and (3) the PUCV-VisibleThermal-Face (PUCV-VTF) database is described for the first time for face recognition, as it was designed for visible-thermal studies in face recognition. The main idea of the fusion method is to maximize face recognition rates using genetic algorithms, which improves recognition rates using visible and thermal descriptors obtained from the newest and best methods currently used in the literature.

The PUCV-VTF database contains thermal and visible images acquired with two different sensors and allows us to study the fusion of these images. The visible images were acquired with a PS3Eye camera, 640 × 480 pixels, while the thermal images were acquired with an FLIR Tau 2 640 camera, with a resolution of 640 × 512 and 14 bits.

This study includes five local matching methods obtained from different comparative studies of visible and thermal methods [7,8,9,10,11,12], such as local binary pattern (LBP) [13], the Weber law descriptor (WLD) [14] and the Gabor jet descriptor (GJD) [11], due to their high performance. In addition, new methods that work in the visible spectrum, such as local derivative patterns (LDP) [15] and HOG descriptors [16], are also included due to the high performance seen in recent studies [15,16,17]. To perform the fusion method and based on results obtained previously [7,8,12], we choose two local descriptor methods, the LDP and LBP descriptors, to perform the combination. Thus, the genetic algorithm combines both descriptors in order to maximize the recognition rate of the entire face recognition system.

The main differences between our proposed fusion method and other fusion studies [18,19,20] are presented in the following paragraph. In [18,19], standard appearance-based methods are used together with genetic algorithms for the analysis and fusion of visible and thermal data. The studies use the classic eigenfaces approach for fitness evaluation, in which it is difficult to describe the face efficiently, unlike with our method, which uses new approaches for face description. In addition, our way of performing the fusion is different, because we use only 64 weights for the fused descriptor against the 100 principal components in [18], for example. In [20], the fusion of visual and thermal images using the discrete wavelet transform (DWT) domain is described. The results of the experiments demonstrate that the fusion method is effective in terms of visual quality compared to conventional visible and thermal approaches, but it depends on glasses being present in the image. In our work, the selected face recognition descriptors are robust to the occlusion produced by glasses [7]. In [21,22], genetic algorithms are not used to perform the fusion; however, those articles use methods based on data fusion and decision fusion. In [21], the fusion algorithm is designed to detect and replace glasses with an eye template in the case of thermal images, and FaceIt, a commercial face recognition software, is used to evaluate the fusion algorithm. In [22], the advantages of combining thermal and visible face recognition are analyzed, and recognition is achieved using a k-nearest neighbor classifier by principal component analysis (PCA). Finally, [23] uses a fusion method based on image fusion using the entropy information with a genetic algorithm. The study does not perform face recognition; however, it shows a way to efficiently combine visible and thermal features to be used in a face recognition system.

The proposed fusion methodology and the current visible-thermal methods will be evaluated using the Equinox database [24] and a new database named PUCV-VTF, which has been created to compare recognition rate results and to validate the fusion method. The Equinox database was used in [7,9,10], and it consists of face images in the visible, long-wave infrared (LWIR), mid-wave infrared (MWIR) and near-wave infrared (NWIR) spectra. For this study, only the visible and LWIR are used with six gallery sets and nine test sets for both spectra. In addition, we will compare the face recognition results obtained using the fusion approaches from [18,19,20] against the proposed methodology using the Equinox database. The PUCV-VTF database consists of face images in the visible and long-wave infrared spectra (thermal images), taken at the same time. We will create five sets, using a frontal set as the gallery and the others as test sets. Specifically, in our experiments, we use the Equinox database to generate the weights with the genetic algorithm for the fusion method, and we then evaluate the fusion method using the PUCV-VTF database.

This paper is structured as follows: Section 2 describes the current visible and thermal methods used in face recognition systems. The proposed methodology of visible-thermal descriptor fusion using a genetic algorithm is explained in Section 3. The visible and thermal databases and the image sensors used in this study are presented in detail in Section 4. Section 5 shows the experiments to obtain the optimal population of the fusion scheme and the validation of the fusion methodology. Finally, the main conclusions are presented in Section 6.

2. VISIBLE-THERMAL METHODS

As was mentioned in the previous section, we selected five current methods used in the literature to generate face recognition systems in the visible and thermal domains. The methods were chosen considering high performance and requirements, such as working in real time, using only one image per person in a gallery for face recognition matching (no training stage) and their high rates in comparative studies [7,8,9,10,11,12].

2.1. LDP Histograms

The local derivative pattern of n-th order was first proposed in [15]. The method obtains micropatterns that use the information contained in directional derivatives of (n − 1) order of the original image. The method uses different α directions to obtain a descriptor: 0°, 45°, 90° and 135° only, since 180°, 225°, 270° and 315° are implicitly contained in the directions used. The micropatterns are obtained using a neighborhood of 8 pixels around a central pixel, taken from the derivative image in the α direction to extract the information using a comparative function. To give a holistic feature to LDP, the image is divided into square regions where histograms with 256 bins are computed. The information taken from all of the directions for every region is concatenated for the final LDP n-th order descriptor of the original image.

2.2. LBP Histograms

The local binary pattern was originally proposed in [13]. It was first designed for the description of textures and was then adapted to face recognition. The method compares the intensity differences between the central pixel and its neighborhood in a 3 × 3 region to generate a binary code. One advantage of this method is its computational efficiency; this is important in image processing in real time. With this method, three levels of locality are obtained: pixel level (information about patterns obtained pixel to pixel), regional level (where groups of pixels give information about small regions of the face) and the holistic level of locality. A global description of the face is obtained concatenating the regional LBP extracted features using histograms by region.

2.3. Histograms of Oriented Gradients

Histograms of oriented gradients, proposed in [16], were used to obtain a human descriptor for pedestrian applications. The HOG algorithm uses the information contained in both the magnitude and orientation of the gradients presented in the original image. The face image is divided into *mxm* regions, where the histograms are calculated with 9 bins using the information on magnitude and orientation. These histograms are concatenated to obtain the final HOG descriptor of the original image, taking half of the precedent region in both directions to overlap the face information.

2.4. Weber Local Descriptor

The Weber local descriptor was first proposed in [14] and is based on the fact that the human perception of a pattern does not depend solely on a stimulus change, but also on its intensity. Specifically, WLD consists of two-dimensional histograms: differential excitations and orientations. The differential excitation component is a function of the radius between two terms: the first term is the relative difference in intensity of the current pixel against its neighbors, and the second term is the intensity of the current pixel itself. The orientation component is the orientation of the gradient of the current pixel. Thus, given an image, both components are used (differential excitation and orientation) to build and concatenate the WLD histogram.

2.5. Gabor Jet Descriptor

Gabor filters are used in face recognition because of their similarity to cell behavior in visual perception and of their robustness to change in illumination. The filters extract characteristics by means of the selection of frequency, orientation and scale. The jets are calculated at different points on a grid, and then, the Borda count method is applied to obtain a ranking and, thus, to generate the recognition. The implementation for this study was based on [11], in which several local feature representations, classification methods and combinations of classifier alternatives were compared. That system uses Gabor jet descriptors as local features, which are uniformly distributed over the images, one wavelength apart. In each grid position of the test and gallery images and at each scale (multiscale analysis), the Gabor jets are compared using normalized inner products. These results are combined using the Borda count method.

To summarize, we used five local descriptor-based methods, LDP, LBP, HOG, WLD and GJD, to perform the face recognition. For LDP, LBP and HOG, we used different divisions of the image to give them a holistic feature in partitions of 20 × 4, 8 × 4 and 16 × 16 regions, in order to find the best partition regions in our analysis. For WLD, we consider the results obtained from [7], which uses only 20 × 4 partitions. The recognition for the local descriptor method is performed using a nearest neighborhood with one of the three following similarity measures: Euclidean distance (EU), histogram intersection (HI) and chi squared (X2). For the case of the Gabor jet descriptor, the image is not partitioned, and the Borda count is used to compute the matching for the face recognition. In addition, the following notation is used to name each method and variation in this study: A_B_C, A describes the name of the method (LBP, HOG, GJD, LDP n-th and WLD), B shows the similarity measure (EU, HI or X2); C denotes the number of regions used (80 for 20 × 4, 32 for 8 × 4 and 256 for 16 × 16). In the case of GJD, only the Borda count is used, and the image is used as a whole, so its notation is simply "GJD".

3. PROPOSED METHOD: VISIBLE-THERMAL FACE DESCRIPTOR FUSION BY GENETIC ALGORITHMS

Genetic algorithms have been used since 1975 by applying Darwin's theory [5,6], and their use is inherently associated with maximizing or minimizing the result of a fit function by means of natural selection and mutation genetics, thus comprising a robust method of finding solutions that is not mathematically guided.

We propose the use of a fusion technique based on genetic algorithms, which combines the most relevant information from visible and thermal images with image descriptors. Our genetic algorithm finds a genetic code, which consists of weights for the visible and thermal descriptors to maximize the recognition rate as an objective function. The fusion method is performed by taking an input pair of test images (visible and thermal) to be compared against a pair in the gallery set, and following this, two face recognition descriptors are applied: the visible descriptor (VD) and the thermal descriptor (TD) (any local descriptor can be chosen). To combine the descriptors, histogram intersection is used as the similarity measure to perform the fusion.Figure 1 shows the fusion method.

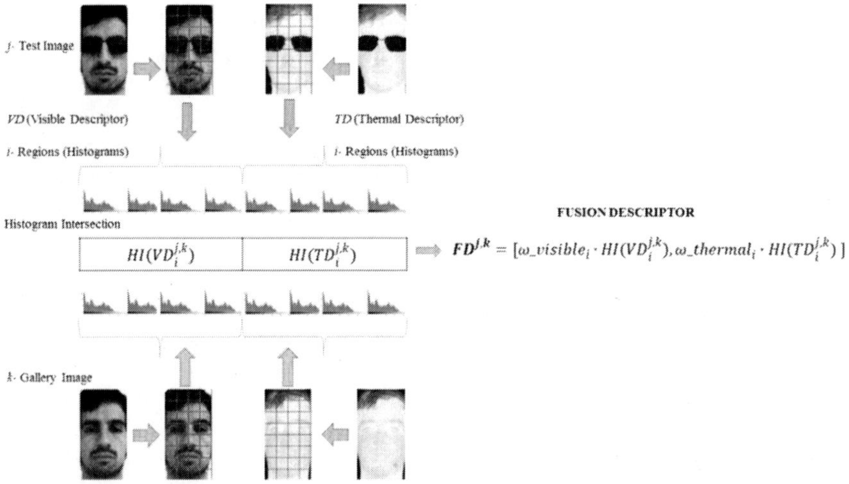

Figure 1. Fusion scheme process.

Combinations of image descriptors were selected based on the results of [7,12,15]; for the visible spectrum, two descriptors were chosen: the third order LDP descriptor (LDP3), with 256 divisions in the image, and the commonly-used LBP; for the thermal case, only the LBP descriptor with 32 regions was used, as it is simple and obtains high results. The fusion method combinations are therefore based on: LDP-LBP and LBP-LBP for the visible and thermal descriptors. In addition, note that the

LDP descriptor for visible image uses 256 regions, and the thermal image uses 32 regions; thus, to perform the fusion, we take groups of 8 sectors for the visible LDP descriptor. Therefore, the length of the genetic code given by the GA is of 64 weights, where the first 32 correspond to visible weights and the following 32 to thermal weights. Once the recognition rate was obtained, our fusion scheme maximized this recognition rate using the weights obtained from the GA process.

3.1. Proposed Genetic Algorithm for Fused Descriptors

Basically, a genetic algorithm (GA) searches for the optimum result of an objective function. First, an initial population of possible solutions has to be created and then evaluated using the objective function (also known as a fitness function) obtaining the fitness value of each individual in the population. After the initial population is established, several steps put the GA into a loop algorithm, which can be stopped by different criteria, such as an expected fitness value or number of iterations. This basic method also involves the selection of individuals as parents, crossover and mutation and finally reinsertion into the population.

The starting point of our GA was the choice of a function to maximize fitness, in our case the recognition rate. The GA assigned weights ω = [ω_*visible* ω_*thermal*] to the partitions of each image (visible and thermal), which represent the genetic code to be optimized. The process considers the face recognition results obtained from each test image *j*, gallery image *k* and the image partitions *i* of the regions of the histogram of the visible descriptor and the thermal descriptor (see Figure 1). Mathematically, the vector fusion descriptor (*FD*) for the test image *j* and the gallery image *k* can be described by the following Equation (1):

$$FD^{j,k} = [\omega\_visible_i \cdot HI(VD_i^{j,k}), \omega\_thermal_i \cdot HI(TD_i^{j,k})] \tag{1}$$

where *i* is the number of regions of the image, *VD* is the visible image descriptor, *TD* is the thermal image descriptor and ω_*visible* and ω_*thermal* are the weights of the population to be optimized. The histogram intersection function *HI(·)* combines the descriptor regions from the test image *j* and gallery image *k* and is given by $HI(S^{j,k}) = \sum_w \min(S_w^j, S_w^k)$.

Once the objective function was chosen and the nature of the genetic code was established, the GA was run with given gallery and test sets, aiming to find the best recognition performance between them using the GA with the following steps.

Initial population:

1. An initial population of 100 genetic codes was randomly made with 64 weights in the interval [0, 1]. The initial population had complementary weights for visible and thermal regions.
2. The genetic code was applied to the descriptor database of gallery and test images using Equation 1. The similarity value was then calculated using $SV^{j,k} = \sum_{i=1}^{64} FD_i^{j,k}$.
3. The face recognition was performed, giving the maximum value of the similarity value (SV) to each test image j, and the recognition rate was then calculated using the correctly-recognized faces. Note that the fitness value of each genetic code is the recognition rate, which was then optimized.

New offspring generation:

4. Two genetic codes were selected as parents, using the roulette method [5,6].
5. The parents were crossed with a probability of 25% at each point, obtaining two off spring (the weights of visible and thermal were no longer complementary).
6. The offspring was mutated at three random points with a probability of 25%.
7. The fitness values of both offspring were calculated as for the initial population (Steps 2–3).
8. If the fitness value of a given offspring was greater than the lowest fitness value in the population, then a random genetic code was replaced, otherwise it was discarded.
9. If the iteration number was under 100,000, Step 4 was repeated.
10. End.

The genetic algorithm process may take a considerable amount of time to obtain the genetic code. However, the GA process is performed only to find the best optimal weights to obtain the best genetic code, similar to an offline stage. After this process, the face recognition is performed with the fusion method using only the best code obtained.

4. VISIBLE AND THERMAL DATABASES

To carry out the fusion analysis, the Equinox database and the PUCV-VTF database were used. A brief description of the databases is given in the following sections. It is important to mention that the PUCV-VTF was created for this research and considers images in both the visible and thermal spectra of 76 individuals and different face conditions, such as emotions, gestures and artefacts (glasses).

4.1. Equinox Database Description

The Equinox database [24] has been used in several studies to compare and evaluate face recognition methods. Thus, it allows direct comparison between methods implemented previously, such as PCA, Independent Components Analysis (ICA), LBP, WLD, GJD, kernel PCA (KPCA) and kernel Fisher linear discriminant (KFLD) [9,10,19,25]. The Equinox database is often used to study the performance of different face recognition algorithms using thermal imaging (8–12 microns). The database consists of several sets of frontal images taken in both visible and infrared spectra at the same time, totaling 18,629 pictures for each spectrum with grey scale with a size of 240 × 320 pixels. The gallery set contains images from 81 different subjects, which were obtained under different lighting (frontal, left lateral and right lateral), three different expressions (smile, frown, surprise), while the test sets have more images per subject (see Table 1). Subjects who use glasses are captured twice, with and without their glasses. The images were processed to be 150 pixels in height and 81 pixels in width, with the eyes horizontally centered, aligned and with a distance between them of 42 pixels, showing only the face of the subjects (see the examples in Figure 2).

For the evaluation of the methods used in this study, 15 sets of images were established with the following criteria (taken from [12]). All subjects have images in each set (visible and thermal); for gallery sets, there is only one image per subject (81 images per set); the test sets have different sizes, as detailed inTable 1. Thus, six gallery sets were established (VA, EA, VF, EF, VL, EL) and nine test sets (VA, EA, VF, EF, VL, EL, VG, EG, RR), see details in Table 1.

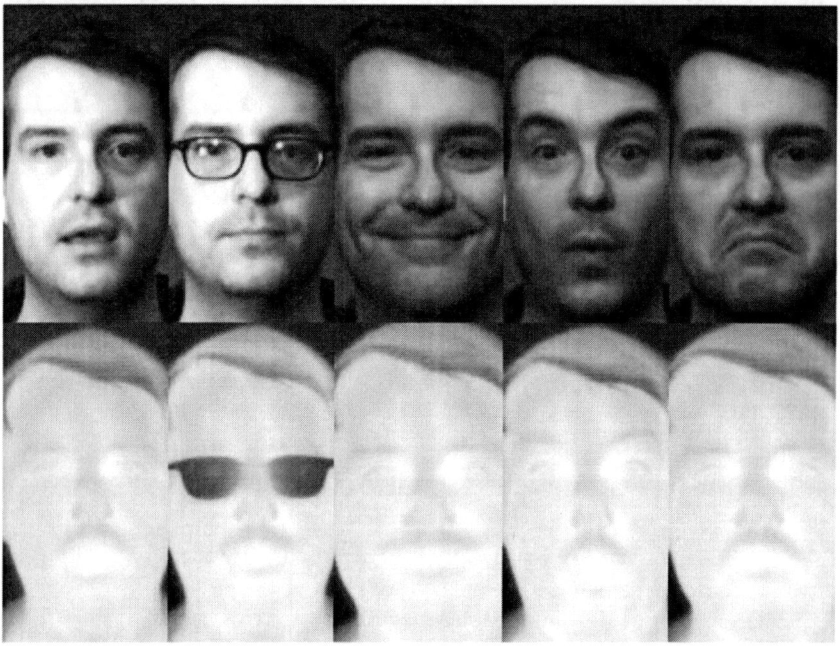

Figure 2. Equinox database examples.

4.2. PUCV-Visible Thermal-Face Database Description

The PUCV-VTF database [26] was developed to contribute to the research community interested in face recognition. The visible images were acquired with a PS3Eye camera [27], with the following specifications: image resolution of 640 × 480 pixels, field of view between 56° and 75° and a frame rate of 120/60 frames per second.

Thermal images were acquired with a FLIR Tau 2 camera [28], with the following specifications: image resolution of 640 × 512 pixels and 14 bits, field of view between 56° and 69°, frame rate of 30 frames per second, thermal sensibility of 50 mK and a spectrum range between 7.5 and 13.5 μm.

Table 1. Equinox Database Description. The quantity of images the test sets is detailed in the table.

Sets	Description	Subjects	Illuminations	Image Number
VA	Vowel Frames	All Subjects	All illuminations	729 images
EA	Expressions frames	All Subjects	All illuminations	729 images
VF	Vowel frames	All Subjects	Frontal illumination	243 images
EF	Expressions frames	All Subjects	Frontal illumination	243 images
VL	Vowel frames	All Subjects	Lateral illumination	486 images
EL	Expressions frames	All Subjects	Lateral illumination	486 images
VG	Vowel frames	Subjects Using Glasses	All illuminations	324 images
EG	Expressions frames	Subjects Using Glasses	All illuminations	324 images
RR	Random 500 frames	Chosen at random	All illuminations	500 images

The physical setup is shown in Figure 3. This consists of a tripod where the two sensors (FLIR Tau 2 and PS3 Eye) are positioned with a separation of 6.5 cm. These cameras are at a height of 115 cm from the floor and at a distance of 91.5 cm from the background. The background is a white wall, which was chosen to avoid thermal and visible changes that might affect the acquisition of the images.

Figure 3. Physical setup of the PUCV- VisibleThermal-Face (VTF) database acquisition system.

The database includes 76 individuals with five subsets, in a total of 12,160 images in both visible and thermal spectra. The sets comprise normal faces (frontal condition), frown, glasses, smile and vowel. Only one image per subject is used to create the sets. The normal set was considered as the gallery, as it is neutral, while the rest were used as test sets. The characteristics of the sets are the following: normal faces indicate a serious pose of the individual; frown indicates anger; glasses are images of the individual using transparent or dark glasses at random; smile shows some grinning; and vowel faces are images of the individuals articulating a vowel of their choice. We use the following notations for each set: Nml for normal, Frn for frown, Gls for glasses, Slm for smile and Vwl for vowel (see the examples in Figure 4).

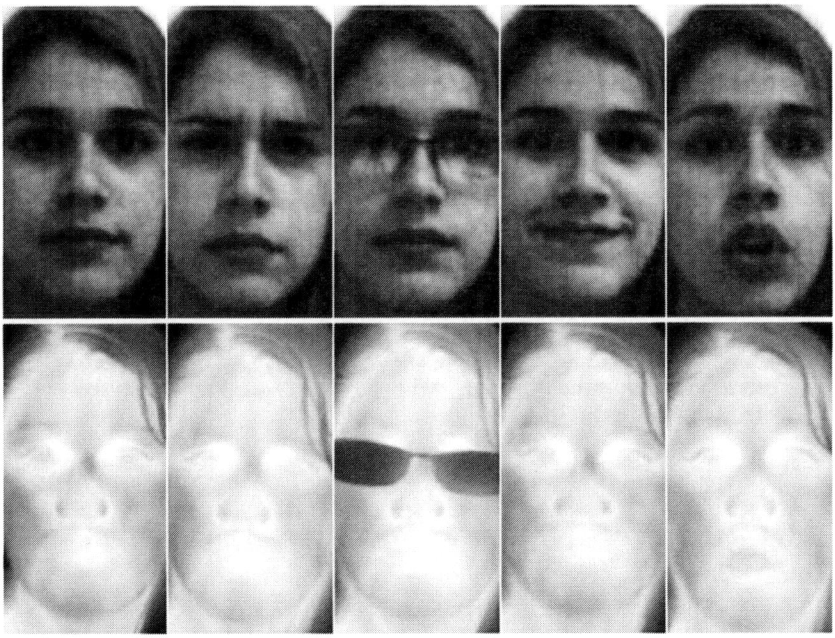

Figure 4. PUCV-VTF database examples.

All of the images were cropped and aligned according to the coordinates of the eyes, which were marked manually. The length and proportion of the distance between the eyes and the borders were made in line with the Equinox database (81 × 150 pixels with 42 pixels between the eyes). The visible and thermal images were normalized using the following Equation (2), which consists of applying a linear mapping to the pixels intensity values in the range [N_{min}, N_{max}]:

$$I_{norm}(i,j) = \frac{I(i,j) - I_{\min}}{I_{\max} - I_{\min}} (N_{max} - N_{min}) + N_{min} \quad \forall (i,j) \in \Omega \tag{2}$$

where I_{min} and I_{max} are the minimum and maximum values in the image: Imin=min(i,j)∈ΩI(i,j) andImax=max(i,j)∈ΩI(i,j). In the experiments, the value range of [0,255] is used.

5. EXPERIMENTS

In order to evaluate the proposed fusion methodology using GA, the face recognition experiments were performed using two databases: Equinox and PUCV-VTF. The first experiment consists of comparing face recognition results from the fusion method against other face recognition methods and the fusion results obtained from [18,19,20], thus obtaining the best configuration of our methodology and the optimal population to achieve high recognition rates. The GA was run in order to obtain the best genetic codes for the fusion method. After the experiment, the optimal population was selected. This is then validated with the PUCV-VTF database, comparing the results with the visible and thermal algorithms used in the literature.

5.1. Experiment 1: Optimal Population in the Equinox Database

The main idea of this experiment is to obtain the best genetic codes with the GA process for the proposed fusion method described in Section 3. The fusion methods used 2 variants composed of the combination of visible and thermal descriptors: LDP-LBP and LBP-LBP descriptors. The fusion method using the LDP3 operator for visible spectrum and LBP descriptor in the thermal is named FD-LDP-LBP, and FD-LBP-LBP is the fusion method using the LBP descriptor for both spectra.

To achieve the best genetic codes, only two subsets (the most representative) of the Equinox database are used to obtain the optimal weights. In the experiment, the GA is trained using the EA set as the gallery set and the RR set as the test set, since this set is random and represents more variability in the images (lighting conditions, expression, artefacts). The GA was applied to the gallery and test sets, finding the best possible population using the proposed fusion methodology following the steps shown in Section 3. The evolution of the genetic algorithm (fitness value) is shown in Figure 5 for both FD-LDP-LBP and FD-LBP-LBP. The optimal final population represented by 100 different individuals was found, achieving the same recognition rate for the gallery and test sets, reaching almost 98% for both fusion descriptor methods. Note that FD-LBP-LBP showed a faster convergence than FD-LDP-LBP to achieve the optimum.

Once the GA process had been performed, the optimal population was tested using the entire Equinox database. In addition, the face recognition rate obtained using the fusion method was compared with the results obtained using the face recognition methods shown in Section 3 and the fusion results obtained from KPCA (kernel PCA) and KFLD (kernel Fisher linear discriminant) [19], wavelets and PCA [18] and discrete wavelet transform (DWT) [20]. The recognition experiments were performed using the face recognition methods applied to the gallery and test sets for the Equinox database. For each gallery and test set, the top one recognition was computed. This consists of obtaining the most similar image as a correct rate from the gallery/test sets. In addition, the average recognition rate of each result from gallery/test pairs of the entire Equinox database was used as the evaluation criteria. This is calculated as the mean of the recognition rate for each set. The results of the fusion approaches obtained from the literature [18,19,20] do not include the same gallery/test sets used in this work, because those studies often removed the case of the subject with glasses. In order to obtain a fair evaluation, the best results reported in each were considered to obtain a comparison criterion with the proposed fusion method.

The results obtained are shown in Table 2 after applying 100 genetic codes obtained from the GA process for both fusion methods. All 100 genetic codes obtained better recognition rate results in comparison with current face recognition methods, even the worst code obtained. Table 2 shows the results from the face recognition method and only two results of the fusion descriptors, *i.e.*, the best variants for FD-LDP-LBP and FD-LBP-LBP, which were chosen to show the effectiveness of the proposed fusion method. In addition, comparing all of the results of the face recognition methods, our proposed methods increase recognition performance to 96.99%. Thus, it is possible to obtain a unique genetic code that ensures the best performance for the face recognition system with the Equinox database using FD-LDP-LBP. In addition, the FD-LBP-LBP also increases performance, achieving 95.34% for recognition and surpassing the classic visible and thermal methods. The fusion methods from the literature obtained high performance rates, achieving ~96% recognition rates; however, these fusion results are less than the results obtained from the proposed fusion method. Previous fusion techniques do not consider difficult cases, such as glasses, which further enhances our results. We believe that the high recognition rates achieved by our fusion methodology are due to the local descriptors used for the face recognition, which are robust to occlusion. Furthermore, the descriptors are not affected by lighting conditions, and the genetic algorithm efficiently combines features for the face description. For these reasons, the next experiments will not consider the evaluation of the classic fusion methods. Note that without using the proposed fusion method, the best variant for thermal face recognition is LBP, obtaining 95.32%, but in the visible case, the best method is LDP3 with only an 88.20% recognition rate.

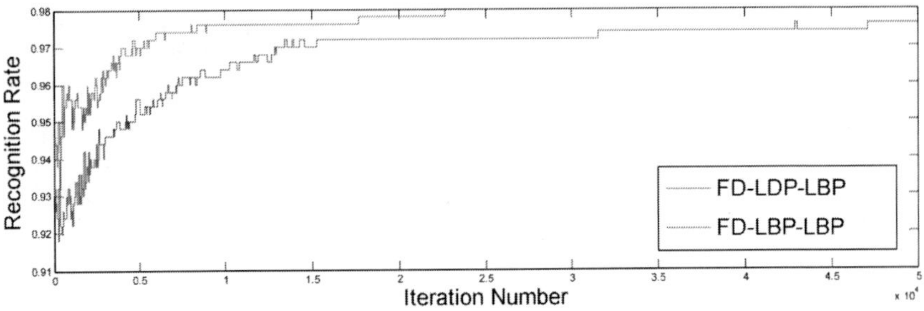

Figure 5. Fitness value evolution for the genetic algorithm. FD, fusion descriptor; LDP, local derivative pattern; LBP, local binary pattern.

Figure 6 presents an analysis of the behavior of the best variants of face recognition methods and the proposed fusion method for all Equinox galleries. The robustness of each method is evaluated by calculating the recognition rate using different conditions, such as expressions, gestures, lighting and glasses, by using different test sets from the Equinox database. Thus, for both spectra, it can be seen that FD-LDP-LBP and FD-LBP-LBP are more robust than other methods to changes in expression and the use of glasses, since they acquire visible and thermal characteristics given by the optimal weights. Additionally, considering the visible spectrum results, most of the visible methods are affected by glasses and illumination, whereas the thermal methods are more robust to expression and illumination conditions, but are affected by glasses, with LDP3 and GJD being the worst methods in this spectrum.

To finish our experiment and as a means of representation, the best genetic codes obtained from FD-LDP-LBP were selected, showing that parts of the visible and thermal image of a human face are useful for face recognition process. For example, given its weight values, the best genetic code found shows that there are areas for face recognition that are more significant in the visible spectrum, and others are more important in the thermal spectrum. Figure 7 shows a face divided into 32 regions, where each region is representing by a color, blue or red, representing the information of the optimal population found with the GA. In addition, inFigure 7, it can be seen that the proposed method chose to give more relevance to the thermal regions (red parts: noose, lips and chin), because these are more invariant to changes in lighting and pose, but in the areas around the forehead and hair, it gives more relevance to visible areas (blue parts), as the thermal imaging has no information. In the next section, the two fusion descriptions, FD-LDP-LBP and FD-LBP-LBP, and the best population are used to analyze another database to validate the performance of our fusion methods.

Table 2. Average recognition rate for the proposed fusion method against all face recognition methods for the Equinox database. Results in bold show the best variation. HI, histogram intersection; WLD, Weber law descriptor; GJD, Gabor jet descriptor; KPCA, kernel PCA; KFLD, kernel Fisher linear discriminant.

Method	Visible (%)	Thermal (%)
LBP_HI_80	81.60	**93.89**
LBP_HI_32	80.11	**95.32**
LDP2_X2_32	83.12	**88.98**
LDP3_HI_256	**88.20**	61.10
LDP3_X2_32	**85.22**	73.30
HOG_X2_256	73.59	**94.03**
WLD_HI_80	78.89	**84.86**
WLD_X2_80	38.05	**85.19**
GJD	**85.35**	70.07
FD-LDP-LBP	**96.99**	
FD-LBP-LBP	95.64	
Wavelets [18]	93.50	
PCA [18]	92.90	
KPCA [19]	82.70	
KFLD [19]	96.30	
Wavelets [20]	96.10	

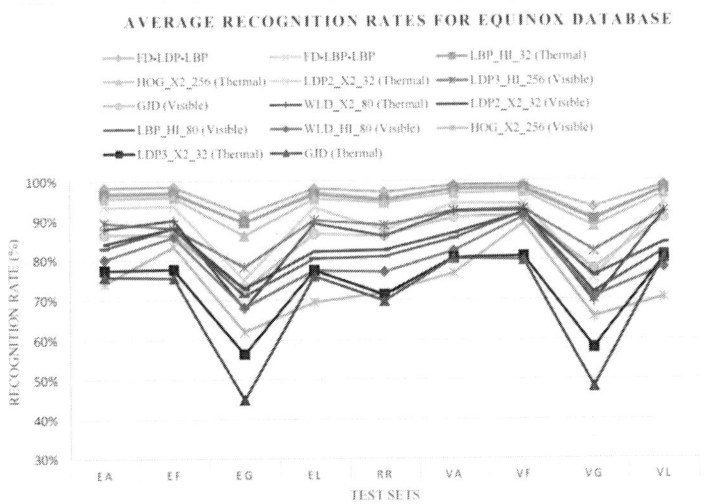

Figure 6. Average recognition rates for the Equinox database. Different test sets against the entire gallery sets.

Figure 7. Optimal genetic code obtained for the GA.

5.2. Experiment 2: Fusion Method Validation Using the PUCV-VTF Database

The second experiment consist of evaluating the PUCV-VTF database with the fusion methodology *versus*the visible and thermal face recognition methods to compare and validate the proposed fusion descriptor. Firstly, the GA was not re-implemented to obtain the best population, since the best population obtained with the Equinox database was used. Secondly, the experiments were conducted using the top one criterion to obtain the recognition rate. Similar to the Equinox database, the different test sets (frown, glasses, smile and vowels) were compared with the gallery (normal) using all of the methods from Section 2 and the proposed fusion method, selecting the best results, which are shown in Table 3. From Table 3, it can be seen that the results are consistent with the Equinox database, where the fusion methodologies achieve the best results and very high recognition rates (99.01%) for FD-LBP-LBP. In addition, one fusion method, FD-LDP-LBP, and the thermal LBP variant, LBP_HI_80, obtain similar results: 98.68%. Note that thermal methods attain high performance in recognition rates with local descriptors, such as LBP, HOG and WLD.

To show the performance of each test set *versus* the gallery set, Table 4 was created. It shows the average recognition rate of all test sets (frown, glasses, smile and vowels) *versus* the gallery set (normal). We can see that for the frown set, the two proposed fusion descriptors obtain the best performance of 98.68%, where, for example, the GJD and LDP variants in the visible and thermal perform very poorly. For the glasses set, most face recognition methods decrease their recognition rates, but the FD-LBP-LBP and the thermal LBP_HI_80 obtain the best results, remaining very robust with the use of this artefact. With the smile set, all methods are very robust to this pose variation in the thermal spectrum, where LBP, WLD, GJD and the proposed fusion methods achieve 100% recognition rates. For the vowel set, something different occurs, because here, the thermal LBP obtains 100% performance and surpasses the proposed fusion descriptor. We believe the reason is that the visible features do not give more information for the proposed fusion descriptor with the vowel set. Finally, the average results of the entire gallery/test sets are compared to identify the best performance, which is seen for the fusion descriptor methodology (FD-LBP-LBP). The result of a 99.01% recognition rate validates the proposed fusion methodology, showing high results for both the Equinox and PUCV-VTF databases.

For both the Equinox and PUCV-VTF databases, the fusion methodology obtains high face recognition performance, achieving the best recognition rates. These results are very encouraging, since, through the use of both spectra, it is possible to combine better features than in the visible and thermal spectra separately. Thus, it can be concluded that the proposed fusion descriptor using a genetic algorithm increases the recognition rates in a visible/thermal face recognition system.

Table 3. Average recognition rate for the proposed fusion method against all face recognition methods for the PUCV-VTF database. Results in bold show the best variation.

Variant	Visible (%)	Thermal (%)
LBP_HI_80	91.45	**98.68**
LBP_X2_32	96.71	**98.03**
LDP2_HI_32	**94.08**	91.78
LDP2_HI_256	91.12	**93.09**
LDP3_X2_32	78.95	**83.22**
LDP3_EU_256	74.01	**86.51**
HOG_EU_256	93.09	**97.04**
HOG_X2_256	92.76	**97.70**
WLD_HI_80	90.46	**97.70**
GJD	80.26	**92.76**
FD-LDP-LBP	98.68	
FD-LBP-LBP	**99.01**	

Table 4. Average recognition rate for the PUCV-VTF database. Different test sets against the entire gallery sets. In bold are the best results per test set.

Variant	Average Recognition Rate-Test Sets				
	Frown	Glasses	Smile	Vowels	Average
LBP_X2_32 (Visible)	96.05	96.05	98.68	96.05	96.71
LBP_HI_80 (Thermal)	96.05	**98.68**	**100.00**	**100.00**	98.68
LDP2_HI_32 (Visible)	94.74	93.42	94.74	93.42	94.08
LDP2_HI_256 (Thermal)	89.47	92.11	97.37	93.42	93.09
LDP3_X2_32 (Visible)	65.79	77.63	86.84	85.53	78.95
LDP3_EU_256 (Thermal)	82.89	78.95	98.68	85.53	86.51
HOG_EU_256 (Visible)	88.16	92.11	96.05	96.05	93.09
HOG_X2_256 (Thermal)	96.05	97.37	98.68	98.68	93.09
WLD_HI_80 (Visible)	85.53	85.53	97.37	93.42	90.46
WLD_HI_80 (Thermal)	96.05	97.37	**100.00**	97.37	97.70
GJD (Visible)	60.53	69.74	97.37	93.42	80.26
GJD (Thermal)	92.11	82.89	**100.00**	96.05	92.76
FD-LDP-LBP	**98.68**	97.37	**100.00**	98.68	98.68
FD-LBP-LBP	**98.68**	**98.68**	**100.00**	98.68	**99.01**

6. CONCLUSIONS

In this study, a new fusion methodology to combine visible and thermal descriptors for face recognition systems was presented. The proposed fusion methodology was evaluated using the standard visible and thermal database, Equinox, and a new database, entitled PUCV-VisibleThermal-Face, created to validate the proposed fusion method and to be used in further research on visible and thermal face recognition. In addition, this study compared the recognition rates of current face recognition methods, such as local derivative patterns, local binary patterns, Weber local descriptors, histograms of oriented gradients and Gabor jet descriptors, and classic fusion techniques used commonly in the literature; against the proposed methodology of fusion descriptors using a genetic algorithm to find the best population and, thus, to achieve an optimal face recognition rate.

To achieve the optimal population for the fusion method, the methodology was tested using a part of the Equinox database, achieving a fitness convergence of 98% recognition rate. Thus, with the optimal population, the fusion descriptor was evaluated using the entire Equinox database and the associated methodology [7,12]. The results show that the two fusion descriptors used (FD-LDP-LBP and FD-LBP-LBP) obtain high performances, surpassing the visible and thermal results of the current face recognition methods and other classic fusion approaches used in the literature. In addition, observing the results obtained by the fusion descriptor, the methodology is very robust to changes in illumination and expression, because it combines thermal and visible information, allowing it to choose optimal face areas where one spectrum is more representative than the other. However, when the individual uses an artefact, such as glasses, and although the fusion descriptor results are superior to the visible/thermal face recognition methods, the performance decreases slightly, since the optimal weights are statically obtained with images that randomly contain glasses and no glasses, as reported in [18,20]. We believe that this can be solved by increasing the variability of the images used to obtain the optimal population (more examples to increase the fitness values/recognition rates).

The fusion methodology was validated using the average recognition rate obtained from the evaluation of current visible/thermal face recognition rates and the PUCV-VTF database. The results show the high performance obtained by the fusion descriptor (FD-LBP-LBP), achieving a recognition rate of 99.01%. This result validates the fusion method, as it shows that the fusion descriptor operates with high recognition rates and is robust to images with different variabilities, such as expressions (frown, smile), gestures and artefacts, such as glasses.

The final conclusion on the fusion descriptors is that combining visible and thermal descriptors increases the performance of the face recognition system. The analysis of the fusion descriptor and the evaluation with the two databases shows that using a genetic algorithm to optimize the recognition rates gives an optimal population and leads to the best performance, due to the selection of the optimal weights and, thus, obtaining the best possible combination of visible and thermal features. We conclude that the proposed fusion method is very robust to variations and artefacts and obtains high recognition rates. However, as a future study, these results could be increased with new techniques based on evolutionary algorithms [29], or with the combination of new face recognition descriptors [30,31], or new texture classification methods published in recent literature.

ACKNOWLEDGMENTS

This research was funded by FONDECYT-Chile Grant 11130466.

AUTHOR CONTRIBUTIONS

Gabriel Hermosilla and Gonzalo Farias conceived and designed the experiments; Francisco Gallardo performed the experiments; Cesar San Martin contributed in the data analysis; All author contributed to write the paper.

CONFLICTS OF INTEREST

The authors declare no conflict of interest.

REFERENCES

1. Akhloufi, M.; Bendada, A.; Batsale, J.C. State of the art in infrared face recognition. *Quant. InfraRed Thermogr. J.* **2008**, *5*, 3–26.
2. FLIR, The Ultimate Infrared Handbook for R&D Professionals. Available online: http://www.flirmedia.com/MMC/THG/Brochures/T559243/T559243_EN.pdf (accessed on 22 July 2015).
3. Chen, X.; Flynn, P.J.; Bowyer, K.W. PCA-based face recognition in infrared imagery: Baseline and comparative studies. In Proceedings of the IEEE International Workshop on Analysis and Modeling of Faces and Gestures (AMFG 2003), Nice, France, 17 October 2003; pp. 127–134.
4. Socolinsky, D.; Selinger, A. Thermal Face Recognition over Time. In Proceedings of the 17th International Conference on Pattern Recognition (ICPR'04), 23–26 August 2004; Volume 4, pp. 187–190.
5. Holland, J.H. *Adaptation in Natural and Artificial Systems*; MIT Press: Cambridge, MA, USA, 1992.
6. Sastry, K.; Goldberg, D.E.; Kendall, G. *Search Methodologies: Introductory Tutorials in Optimization, Search and Decision Support Techniques*; Burke, E., Kendall, G., Eds.; Springer: Berlin, Germany, 2005.
7. Hermosilla, G.; Ruiz-del-Solar, J.; Verschae, R.; Correa, M. A Comparative Study of Thermal Face Recognition Methods in Unconstrained Environments. *Pattern Recognit.* **2012**, *45*, 2445–2459.
8. Ruiz-del-Solar, J.; Verschae, R.; Correa, M. Recognition of Faces in Unconstrained Environments: A Comparative Study. *EURASIP J. Adv. Signal Process.* **2009**, *2009*, 1–19.
9. Selinger, A.; Socolinsky, D. Appearance-Based Facial Recognition Using Visible and Thermal Imagery: A Comparative Study. Technical Report; Equinox Corporation: New York, NY, USA, 2001.
10. Socolinsky, D.; Selinger, A. A comparative analysis of face recognition performance with visible and thermal infrared imagery. In Proceedings of the 16th International Conference on

Pattern Recognition (ICPR), Quebec, QC, Canada, 11–15 August 2002; pp. 217–222.

11. Zou, J.; Ji, Q.; Nagy, G. A Comparative Study of Local Matching Approach for Face Recognition. *IEEE Trans. Image Process.* **2007**, *16*, 2617–2628.

12. Hermosilla, G.; Farias, G.; Vargas, H.; Gallardo, F.; San-Martin, C. Thermal Face Recognition Using Local Patterns. In *Progress in Pattern Recognition, Image Analysis, Computer Vision, and Applications*; Lecture Notes in Computer Science; Springer: Puerto Vallarta, Jalisco, Mexico, 2014; Volume 8827, pp. 486–497.

13. Ahonen, T.; Hadid, A.; Pietikainen, M. Face Description with Local Binary Patterns: Application to Face Recognition. *IEEE Trans. Pattern Anal. Mach. Intell.* **2006**, *28*, 2037–2041.

14. Chen, J.; Shan, S.; He, C.; Zhao, G.; Pietikäinen, M.; Chen, C.X.; Gao, W. WLD: A Robust Local Image Descriptor. *IEEE Trans. Pattern Anal. Mach. Intell.* **2010**, *32*, 1705–1720.

15. Zhang, B.; Gao, Y.; Zhao, S.; Liu, J. Local Derivative Pattern Versus Local Binary Pattern: Face Recognition with High-Order Local Pattern Descriptor. *IEEE Trans. Image Process.* **2010**, *19*, 533–544.

16. Dalal, N.; Triggs, B. Histograms of Oriented Gradients for Human Detection. In Proceedings of the 2005 IEEE Computer Society Conference on Computer Vision and Pattern Recognition (CVPR'05), San Diego, CA, USA, 25 June 2005; Volume 1, pp. 886–893.

17. Deniz, O.; Bueno, G.; Salido, J.; de la Torre, F. Face recognition using histograms of oriented gradients.*Pattern Recognit. Lett.* **2011**, *32*, 1598–1603.

18. Bebis, G.; Gyaourova, A.; Singh, S.; Pavlidis, I. Face recognition by fusing thermal infrared and visible imagery. *Image Vis. Comput.* **2006**, *24*, 727–742.

19. Desa, S.; Hati, S. IR and Visible Face Recognition using Fusion of Kernel Based Features. In Proceedings of the 19th International Conference on Pattern Recognition (ICPR 2008), Tampa, FL, USA, 8–11 December 2008; pp. 1–4.

20. Kwon, O.K.; Kong, S.G. Multiscale fusion of visual and thermal images for robust face recognition. In Proceedings of the IEEE

International Conference Computational Intelligence for Homeland Security and Personal Safety, Orlando, FL, USA, 31 March–1 April 2005; Volume IV, pp. 112–116.

21. Abidi, B.; Huq, S.; Abidi, M. Fusion of visual, thermal, and range as a solution to illumination and pose restrictions in face recognition. In Proceedings of the IEEE International Carnahan Conference on Security Technology, Albuquerque, NM, USA, 11–14 October 2004; pp. 325–330.

22. Pop, F.M.; Gordan, M.; Florea, C.; Vlaicu, A. Fusion based approach for thermal and visible face recognition under pose and expresivity variation. In Proceedings of the Roedunet International Conference (RoEduNet), Sibiu, Romania, 24–26 June 2010; pp. 61–66.

23. Erkanli, S.; Li, J.; Oguslu, E. Fusion of Visual and Thermal Images Using Genetic Algorithms. In *Bio-Inspired Computational Algorithms and Their Applications*; Gao, S., Ed.; InTech: Rijeka, Croatia, 2012.

24. Equinox Corporation. Available online: http://www.equinoxsensors.com/products/HID.html(accessed on 1 March 2011).

25. Socolinsky, D.; Wolff, L.; Neuheisel, J.; Eveland, C. Illumination invariant face recognition using thermal infrared imagery. In Proceedings of the IEEE Conference on Computer Vision and Pattern Recognition, Kauai, HI, USA, 8–14 December, 2001; Volume 1, pp. I-527–I-534.

26. PUCV-VTF database. Available online: http://goo.gl/5bWnYp (accessed on 22 July 2015).

27. PS3 eye camera. Available online: http://us.playstation.com /ps3/accessories/playstation-eye-camera-ps3.html (accessed on 5 March 2015).

28. IR Camera FLIR TAU 2. Available online: http://www.flir.com /cvs/cores/uncooled/products/tau(accessed on 22 July 2015).

29. Jothi, J.A.A.; Rajam, V.M.A. Segmentation of Nuclei from Breast Histopathology Images Using PSO-based Otsu's Multilevel Thresholding. In *Artificial Intelligence and Evolutionary Algorithms in Engineering Systems*; Advances in Intelligent Systems and Computing; Springer: Kumaracoil, India, 2015; Volume 325, pp. 835–843.

30. Liu, H.; Hsu, S.; Huang, C. Single-sample-per-person-based face recognition using fast Discriminative Multi-manifold Analysis. In Proceedings of the Asia-Pacific Signal and Information Processing Association, 2014 Annual Summit and Conference (APSIPA), Siem Reap, Cambodia, 9–12 December 2014; pp. 1–9.

31. Hu, S.; Choi, J.; Chan, A.; Schwartz, W. Thermal-to-visible face recognition using partial least squares.*J. Opt. Soc. Am.* **2015**, *32*, 431–442.

6

CHAPTER

AN INTERVAL-VALUED APPROACH TO BUSINESS PROCESS SIMULATION BASED ON GENETIC ALGORITHMS AND THE BPMN

Mario G.C.A. Cimino * and Gigliola Vaglini

Department of Information Engineering, University of Pisa, Largo Lucio Lazzarino 1, Pisa 56122, Italy

ABSTRACT

Simulating organizational processes characterized by interacting human activities, resources, business rules and constraints, is a challenging task, because of the inherent uncertainty, inaccuracy, variability and dynamicity. With regard to this problem, currently available business process simulation (BPS) methods and tools are unable to efficiently capture the process behavior along its lifecycle. In this paper, a novel approach of BPS is presented. To build and manage simulation models according to the proposed approach, a simulation system is designed, developed and tested on pilot scenarios, as well as on real-world processes. The proposed approach exploits interval-valued data to represent model parameters, in place of conventional single-valued or probability-valued parameters.

Indeed, an interval-valued parameter is comprehensive; it is the easiest to understand and express and the simplest to process, among multi-valued representations. In order to compute the interval-valued output of the system, a genetic algorithm is used. The resulting process model allows forming mappings at different levels of detail and, therefore, at different model resolutions. The system has been developed as an extension of a publicly available simulation engine, based on the Business Process Model and Notation (BPMN) standard.

Keywords: BPMN; business process optimization; business process simulation; genetic algorithms; interval-valued parameter

1. INTRODUCTION AND MOTIVATION

Inadequate process design increases inefficiency and yields ineffectiveness. As a consequence, internally observable circumstances, such as backlog, long response times, poorly performing activities, unbalanced utilization of resources and low service levels, may produce external events of interest for the end user, such as increasing damage claims, angry customers and loss of goodwill. Hence, it is important to analyze business processes (BPs) before starting production and also while producing, for finding design flaws and for diagnosis/decision support.

BPs can be seen as a set of partially ordered activities intended to reach a goal. An interesting point from a scientific perspective is to separate the management of the BPs from the technological applications and methods developed for specific domains. Since the 1980s, researchers have been pursuing standard and unified ways of representing and measuring processes [1].

This section is structured into three subsections: modeling, parameterization and simulation of BPs.

1.1. Business Process Modeling

BP modeling is an established way of documenting BPs. A BP model is a generic description of a class of BPs. BP models describe how BP instances have to be carried out. They highlight certain aspects and omit others. A conceptual BP model is independent of a particular technology or organizational environment, whereas an executable BP model is specialized to a particular environment. The development of business process models is very labor-intensive [2]. There is a multitude of languages to support BP modeling, such as textual language (e.g., formal or natural language) and visual language (e.g., flow chart), and there are several representation standards. The interested reader is referred to [3] for a comparative analysis of such languages. Business Process Model and Notation (BPMN) [4] has been an Object Management Group (OMG) standard since 2005, aimed at providing a notation readily understandable by all business stakeholders. The BPMN structure is similar to well-known flow charts and activity diagrams. However, BPMN provides support to represent the most common control-flow patterns occurring when defining process models [5]. In this paper, BPMN is considered a reference standard in BP modeling.

In the process-based approach, quantifiable measurements must be defined, so-called key performance indicators (KPIs). KPIs can be related to a marketing-based perspective (e.g., customer satisfaction), to internal quality (non-compliance) and efficiency (cost, duration). KPIs are created on the basis of business objectives and are the detailed specifications used to track business objectives. A KPI is then associated with a specific process and is generally represented by a numeric value. A KPI may have a target and allowable margins, or lower and upper limits, forming a range of performances that the process should achieve. KPIs can be made up of one or more metrics. The calculated results of the metrics during process monitoring are used to determine whether the target of the KPI has been met. For example, waiting time, processing time, cycle time, process cost and resource utilization are commonly used KPIs [6,7]. To choose the right KPIs of a process requires a good understanding of what is important to the organization [8].

1.2. Business Process Parameterization

Business process simulation (BPS) [9] can be of great help in the quantitative measurement of KPIs. To simulate a process, it must first be defined as a workflow, showing the flow of work from trigger to result, with its main paths (scenarios) [2]. To simulate scenarios of a defined workflow, the availability of data input and parameters is critical. For instance, each case to be simulated must contain a number of process instances and their arrival rate. Moreover, the number of available resources must be declared for the process instances defined in the model. Furthermore, execution data has to be defined. For instance: the duration of each task, the branching proportion of each alternative flow, the number of resources needed by each task execution, and so on. Cost and other quality parameters can also be defined.

There are several possible outcomes of a BPS tool. The output can be a detailed process log that can be analyzed with a process mining framework [10], a set of benchmarks with related diagrams, a set of KPIs values, *etc.* Some simulation tools provide also support for animation (see [11]) and other criteria. The interested reader is referred to [12] for an analysis of such criteria at the application level.

This paper is focused on an important aspect of a BPS tool, *i.e.*, the ability to capture the inherent uncertainty, inaccuracy, variability and dynamicity of a process [13,14]. Most business systems contain variability in the demand of the system (e.g., customer arrivals), as well as in the duration of processes (e.g., customer service times) within the system. In such cases, the use of average values does not provide reliable performance information. To model such variability is crucial when comparing as-is and to-be models, or in customer-based systems, where performance should not drop below a certain level. A great source of uncertainty is also related to the modeling of human resources [15].

A response to the above limits is to make more detailed models. However, a simulation model should have the right level of detail, and adding further detail does not always solve the problem [9]. Hence, it is very difficult to calibrate BPS models. For this reason, BPS tools permit the incorporation of statistical distributions, to provide an indication of both the range and variability of the performance of the process. Hence, task execution times and process arrival rates can be defined by an average value plus some distribution information, commonly standard, uniform or exponential distribution. In addition, BPS tools can incorporate statistical distributions to model non-deterministic decision flows.

A different representation dealing with uncertainty is the fuzzy-valued variable. In contrast with probability, which is an indicator of the frequency or likelihood that an element is in a class, fuzzy set theory quantifies the similarity of an element to a class by means of a membership function valued in the real unit interval [0,1]. While single-valued and probability-valued variables take numerical values, in fuzzy-valued variables, non-numeric values are often used to facilitate the expression of facts. For instance, variable, such as temperature, may have a value, such as cold, warm and hot, represented by three related functions mapping a temperature scale onto similarity values. Thus, a temperature value, such as 25 °C, can belong to both cold and warm with degree 0.2 and 0.8, respectively. In practice, fuzzy sets generalize classical sets whose membership degrees only take values of zero or one.

A special case of a classical bivalent set is the interval, *i.e.*, a set of real numbers with the property that any number that lies between two numbers in the set is also included in the set. An interval-valued parameteris $[x, x]$ is the simplest representation of uncertainty, asserting that measured values in a set of process instances range from x to x. As an example, consider the temperature of a chemical process, which may have values ranging in [19,27]°C. With an interval, no information about the distribution of the measured values is provided, which means that the shape of the probability density function is unknown. In contrast, a uniform probability distribution, $U[x, x]$, asserts that the probability density function is known to be $1 / (x - x)$ for $x \leq x \leq x$ and zero for $x < x$ or $x > x$, which is much more detailed information on the outcomes of a process.

Depending on the scale of modeling, a certain amount of important data about the processes needs to be collected and analyzed in order to be incorporated in a model. Data is usually collected through discussions with experts and particularly with people involved in the processes to be modeled, through observation of the existing processes and studying the documentation about processes. Unfortunately, process traceability is, in general, a very difficult task [16]. In practice, in many cases, the available data is inaccurate and not sufficiently precise to parameterize the model, and there are no sufficient sample values to calculate a probability distribution. Indeed, data gathering is one of the major barriers to BPS methods. In the field of data-based fuzzy modeling, the construction of fuzzy sets is often difficult, even for process experts, and practicable for applications with a small number of variables. In the literature for the design of fuzzy sets, different approaches can be found, as the interpolation between pairs of observation values and given membership values or the data-based extraction via a clustering algorithm [17,18]. Similarly, to determine the most fitting probability distribution with an estimation of the parameters on the basis of observed data is an expensive task. Indeed, even to select the simplest model (e.g., the uniform probability distribution) in a group of candidate models needs statistical methods based on a number of sample data.

In contrast, interval-valued data refer to the data observed as a range instead of as a distribution of values. Interval data require simply the determination of the lower and upper bounds, which often can be easily asked from people who are involved in the activity. Thus, with respect to probability- and fuzziness-valued data, the modeling of interval-valued data requires less data collecting and processing.

1.3. Business Process Simulation

Statistical and interval-based simulation engines are very different in design. A statistical simulation engine evaluates the model on some stochastically extracted input data. This method does not produce an exact, reliable and deterministic output, as multiple executions of the method would produce different results. Moreover, a statistical engine is not scalable, because it requires an exploding number of evaluation points for increasing parameters of the model. In contrast, in a business model with interval-valued parameters, the output interval is determined by the highest and lowest KPI values that can be provided by giving to parameters all possible values within the related intervals. As an example, consider the temperature and the relative humidity of a chemical process, which may have values ranging in [19,27] °C and [74,81] %, respectively, producing a solidification time ranging in [187,231] min.

The problem of processing an output interval can actually be split into two sub-problems; for a given prefixed input: (i) to find the highest KPI value in the output; and (ii) to find the lowest KPI value in the output, under the constraints established by the intervals-valued parameters. As a matter of fact, each of these sub-problems is based on an optimization problem: (i) a maximization problem; and (ii) a minimization problem; respectively.

In the literature, different solution methods have been applied for the optimization of business processes [19]. Basically, search is the solving method for such problems whose solution cannot be determined *a priori*via a sequence of steps. Search can be performed with either heuristic or blind strategies, depending on the use or not of domain knowledge to guide the search, respectively. Further, a search strategy can be more or less based on exploitation and exploration, *i.e.*, using the best available solution for possible improvement and exploring the search space with new solutions, respectively. For instance, Hill-climbing and random search are two examples of strategies based on exploitation and exploration, respectively [20]. A genetic algorithm (GA) is a general-purpose search method, based on principles of natural biological evolution, which combines directed and stochastic strategy to make a balance between exploration and exploitation. GAs have been successfully applied to many industrial engineering problems that are impracticable for conventional optimization methods [21,22]. In a conventional optimization method, a single solution is increasingly improved through iterations, via problem-dependent progress strategies.

In this paper, we propose a novel approach of BPS consisting of the use of interval-valued parameters as an alternative to conventional single-valued or probability-valued parameters. With regard to this problem, we designed and developed a simulation system, called Interval Bimp (IBimp), whose optimization module is based on a GA. The IBimp system has been tested on pilot scenarios, as well as on real-world scenarios. The proposed simulator has been publicly released as an extension of Bimp, see [23], a publicly available simulation engine based on the BPMN standard. Unlike the approaches proposed in the literature, our interval-based simulator can be used in two ways: (i) as a generator of characteristic curves of models subject to uncertainty associated with their input parameters; (i) as a tool to evaluate the worst/best cases within a constrained operating area.

The paper is structured as follows. Section 2 introduces the background notions required to understand the proposed approach. Section 3 discusses existing work in BPS, with attention to parameterization approaches. Section 4 provides specification and design details on our interval-valued BPS approach. Architectural implementation details are provided in Section 5. In Section 6, five scenarios are introduced to illustrate the basic concepts of our approach. Section 7 covers conclusion and future works.

2. BACKGROUND

The importance of conceptual modeling is largely recognized in the literature [24]. Modeling is a learning process allowing business analysts to make clear requirements, express domain knowledge and provide rough solutions. BP models should have a formal foundation. Indeed, formal models do not allow ambiguity and increase the potential for analysis [25]. In addition, a BP model should be easily understood by all the involved stakeholders. This important requirement can be achieved through the use of visual notations. Once consensus among stakeholders has been reached, the BP model can be deployed on BP management platforms and, if a formal language was used, its behavior can be unambiguously implemented by vendors. BP models may involve core business or complex business transactions. Hence, the analysis of BP models is important to reduce the risk of costly corrections at a later stage, as well as to investigate ways of improving processes. Visual models act as communication channels between business managers and technicians and provide documentation to manage post-project activities.

The strength of BPMN resides in two important aspects: (i) simplicity, which is due to the abstraction level provided by the standard; and (ii) the possibility of being automatically translated into a business execution language and, then, to generate a machine-readable prototype of business processes. BPMN was developed with a solid mathematical foundation provided by exploiting process calculus theory [26]. This theory is an essential requirement for a good business process modeling language, in order to automate execution and to easily provide proofs of general consistency properties, as is widely recognized in the literature [10,27]. Hence, BPMN was conceived of with the specific intention of creating a bridge from the business perspective to the technical perspective about processes [26].

To describe BPs, BPMN offers the BP diagram. Figure 1 shows the basic elements of a BP diagram that are supported by the Bimp simulator: event, gateway (split node) and merge node, task and activity [4]. Key concepts are briefly defined in the following. Events are representations of something that can happen during the BP; a business flow is activated by a start event and terminated by an end event, while intermediate events can occur anywhere within the flow. Business activities can be atomic (tasks) or compound (processes and sub-processes, as a connection of tasks); gateways represent decision points to control the business flow. The sequence flow is represented by solid arrows and shows the order of execution of activities in the BP.

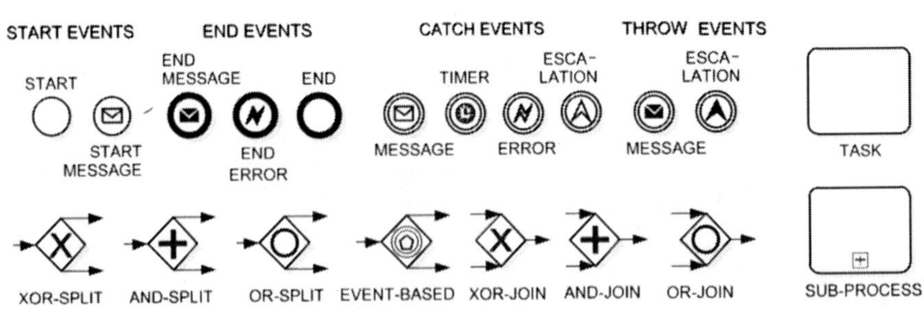

Figure 1. Business Process Model and Notation (BPMN) elements supported by the Bimp simulator.

To define the BPMN process model, we employ the concept of a token traversing the sequence flow and passing through the elements in the process. A token is a theoretical concept that is used to define the behavior of a process that is being performed. During the simulation, any

BPMN element in the process model has its own state per process instance. In Figure 2, a UML (Unified Modeling Language) state diagram shows the lifecycle of a BPMN element supported by the Bimp simulator. More specifically, an element becomes enabled when it has been selected by the simulation engine as an element to be handled. When a resource is assigned to a task, it is said that the task has been started. Elements other than tasks do not have the state, started, because resources can be assigned only to tasks. If the element gets completed without interruptions, the element is said to be completed. Otherwise, it changes to withdrawn.

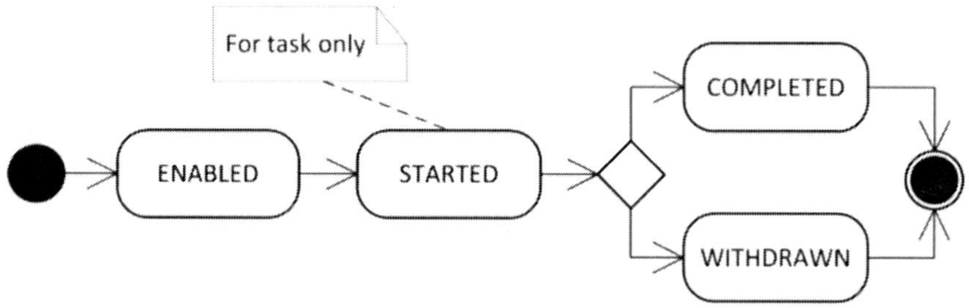

Figure 2. A UML (Unified Modeling Language) state diagram representing the lifecycle of a BPMN element supported by the Bimp simulator.

In general, a BP diagram is organized in terms of interacting pools (partners), represented as labeled rectangular containers. A pool can be divided into lanes, represented as internal labeled rectangular containers, to express a detailed categorization of activities. Finally, message flows, represented as dashed white arrow, express messages exchanged between business entities. An example of a BP diagram is shown in Figure 3. It represents the macro processes of bag manufacturing in a leather workshop. More specifically, the start event in the leather workshop pool indicates where the process starts. Then, a cutting activity is performed on a leather sheet. At this point, there are two possibilities: either the bag components are prepared internally, via a preparing components activity, or externally, by means of outsourcing. In the latter case, an outsourcing management activity is in charge of appointing a third party to the task, which will be carried out by a craftsman. Subsequently, an assembling activity can be either performed internally or externally, as done with the previous activity. Finally, the quality check and packing activity is carried out, and final products are made. The end event indicates where the process ends.

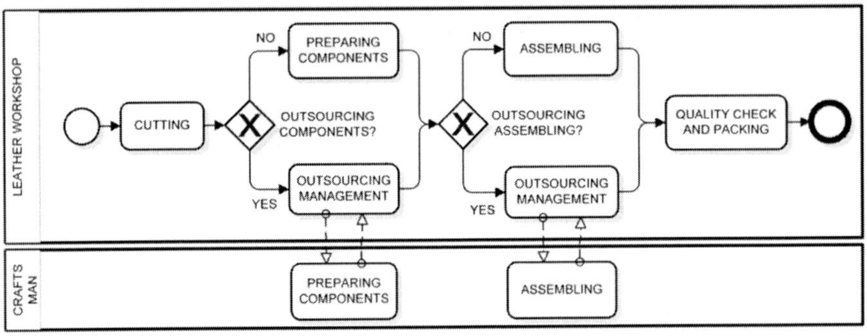

Figure 3. A BPMN process diagram representing the macro processes of bag manufacturing in a leather workshop.

3. RELATED WORK

3.1. Business Process Management Lifecycle

The BP management initiative aims at making processes visible and measurable, to continuously improve them. The BP management lifecycle is made of iterative improvements via identification, discovery, analysis, improvement, implementation and monitoring-and-controlling phases [4,8,28,29]. The identification phase determines an overall view of the process, whereas the discovery and analysis phases take an as-is view of the process and determine the issues of the process to be improved, respectively. The improvement and the implementation phases identify changes to address the above issues and make the to-be model effective. Finally, in the monitoring and control phase, process instances are inspected for the next iteration of the BP management lifecycle. Thus, managing a process requires a continuous effort, avoiding degradation. This is why the phases in the BP management lifecycle should be circular, *i.e.*, the output of the monitoring-and-control phase feeds back into the discovery, analysis and improvement phases.

BP analysis, which is the focus of this paper, includes a rather broad meaning encompassing simulation, diagnosis, verification and performance analysis of BPs [1,28]. BP analysis should offer both supply-chain-level and company-level views of the processes, paying attention to roles and responsibilities. BP analysis tools should be usable by organizational managers rather than by specialists. Three types of analysis can be considered in the field [30]: (i) diagrammatic analysis, related to visual workflow-based models, which enable high-level specification of system interactions, improve system integration and support performance analysis [31]; (ii) formal/mathematical analysis, which is aimed at setting performance indicators related to the attainment of strategic goals [9]; and (iii) language-based analysis, which enables algorithmic analysis for validation, verification and performance evaluation [32]. In particular, performance evaluation aims at describing, analyzing and optimizing the dynamic, time-dependent behavior of systems [30]. The performance level focuses on evaluating the ability of the workflow to meet requirements with respect to some KPIs target values.

A performance measure or KPI is a quantity that can be unambiguously determined for a given BP; for example, several costs, such as the cost of production, the cost of delivery and the cost of human resources. Further refinement can be made via an aggregation function, such as count, average, variance, percentile, minimum, maximum or ratios; for instance, the average delivery cost per item. In general, time, cost and quality are basic dimensions for developing KPIs. The definition of performance measures is tightly connected to the definition of performance objectives. An example of a performance objective is "customers should be served in less than 30 min". A related example of a performance measure with an aggregation function is "the percentage of customers served in less than 30 min". A more refined objective based on this performance measure is "the percentage of customer served in less than 30 min should be higher than 99%".

BPS facilitates process diagnosis (*i.e.*, analysis) in the sense that by simulating real-world cases, what-if analyses can be carried out. Simulation is a very flexible technique to obtain an assessment of the current process performance and/or to formulate hypotheses about possible process redesign. BPS assists decision-making via tools allowing the current behavior of a system to be analyzed and understood. BPS also helps predict the performance of the system under a number of scenarios determined by the decision-maker.

3.2. Business Process Simulation

Modern simulation packages allow for both the visualization and performance analysis of a given process and are frequently used to evaluate the dynamic behavior of alternative designs. Visualization and a graphical user interface are important in making the simulation process more user-friendly. The main advantage of simulation-based analysis is that it can predict process performance using a number of quantitative measures [13]. As such, it provides a means of evaluating the execution of the business process to determine inefficient behavior. Thus, business execution data can feed simulation tools that exploit mathematical models for the purpose of business process optimization and redesign. Dynamic process models can enable the analysis of alternative process scenarios through simulation by providing quantitative process metrics, such as cost, cycle, time, service-ability and resource utilization [30]. These metrics form the basis for evaluating alternatives and for selecting the most promising scenario for implementation.

However, BPS has some disadvantages. Some authors report the large costs involved and the large amount of time to build a simulation model, due to the complexity and knowledge required. The main reason is that business processes involves human-based activities, which are characterized by uncertainty, inaccuracy, variability and dynamicity. Even though simulation is well-known for its ability to assist in long-term planning and strategic decision-making, it has not been considered a main stream technique for operational decision-making, due to the difficulty of obtaining real-time data in a timely manner to set up the simulation experiments [33]. Reijers *et al.* [22] introduced the concept of "short-term simulation". They went on to experiment with short-term simulations from a "current" system state to analyze the transient behavior of the system, rather than its steady-state behavior [34]. In [35], a short-term analysis architecture was designed, in the context of widely-used, off-the-shelf workflow tools and without a specific focus on resourcing. An example of a typical question for a simulation scenario might be "How long will it take to process?" Using conventional tools for BPS, it is possible to answer this question with an average duration, assuming some "typical" knowledge regarding the available resources. Another question might be "What are the consequences of adding some additional resources of a given type to assist in processing?" Again, the question cannot be answered with precision using the "average" results produced by a conventional simulation.

Basic performance measures in BPS are cycle time, process instances count, resource utilization, and activity cost [33]. Cycle time represents the total time a running process instance spends traversing a process, including value-added process time, waiting time, movement time, *etc.* The calculation of minimum, average and maximum cycle time based on all running process instances is a fundamental output of a BPS. The process instance count includes the total number of completed or running process instances. During a simulation, resources change their states from busy to idle, from unavailable to reserved. Current resource utilization, for a given type or resource, defines the percentage of a resource type that has been spent in each state. The availability and assignment of resources dictate the allocation of resources to activities. Hence, resource utilization provides useful indexes to measure and analyze under-utilization or over-utilization of resources. A resource can be defined by the number of available units, usage costs (*i.e.*, a monetary value), setup costs and fixed costs. Cost can include the variable cost related to the duration (e.g., hourly wages of the involved human resources), as well as a fixed additional cost (e.g., shipping cost). When an activity is defined, it is also defined by the resources required to perform it, the duration for the activity and the process instances that it processes. During simulation, the BPS tool keeps track of the time each process instance spends in an activity and the time each resource is assigned to that activity. Hence, activity cost calculations provide a realistic way for measuring and analyzing the costs of activities. BPS tools allow a detailed breakdown of activity costs by resource or process instance type, as well as aggregated process costs.

Table 1 summarizes some important simulation parameters. A simulation is divided into scenarios, and each scenario must contain a well-defined path (from a start event to an end event), with a number of process instances and their arrival rate. Moreover, the available resources must be declared for the process instances defined in the model. Furthermore, element execution data has to be defined; for instance, the duration of each task, the branching proportion of each outgoing flow of a XOR and OR gateway, the resources needed by each task. Cost and other quality parameters can also be defined.

There are several possible outputs of a BPS tool, as represented in Figure 4:

(i)A detailed process log of each process instance, which can be analyzed with a process mining framework [10]. Logs are finite sets of transactions involving some process items, such as customers and products. For example: (1) Jane Doe buys pdt1 and pdt2; (2) John Doe buys pdt1; (3) Foo buys pdt2, pdt3 and pdt4. Logs are usually represented in a format used by the majority of process-mining tools, known as MXML (Mining XML) [36].

(ii)A set of benchmarks with some diagrams. Benchmarking is a popular technique that a company can use to compare its performance with other best-in-class performing companies in similar industries [37,38]. For example, a comparative plot of the compliance to international warranty requirements in a specific industrial sector.

(iii)A set of KPIs values, as exemplified in Table 2.

Unfortunately, for a large number of practical problems, to gather knowledge, modeling is a difficult, time-consuming, expensive or impossible task. For example, let us consider chemical or food industries, biotechnology, ecology, finance, sociology systems, and so on. The stochastic approach is a traditional way of dealing with uncertainty. However, it has been recognized that not all types of uncertainty can be dealt with within the stochastic framework. Various alternative approaches have been proposed [39], such as fuzzy logic and set theory. Figure 5 shows some different representations of a simulation parameter. Here, in Figure 5a, the single-valued representation is shown. The possible (true) value is a unique value, and all other values are false (Boolean membership). With this kind of parameter, any process instance performs the same way (determinism). Figure 5b shows the interval-valued representation. The true values are fully established, and all other values are false (Boolean membership). Again, with this kind of parameter, any process instance performs the same way (determinism).

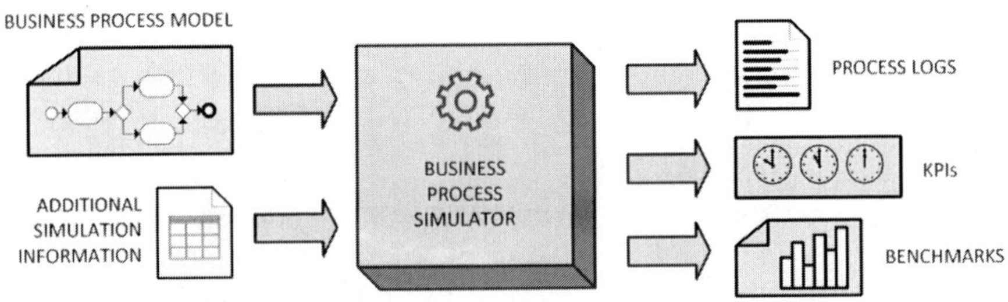

Figure 4. Representation of the input and output data of a business process simulator.

Table 1. Basic simulation parameters.

Parameter	Description
task duration	the time taken by the task to complete
branching proportion	the percentage of process instances for each outgoing flow of an XOR/OR gateway
resource allocation	the resources needed by each task
task cost	a monetary value of the task execution instance
available resources	the number of pools, lanes, actors or role available for tasks
number of instances	the number of running process instances for the scenario
arrival rate	the time interval between the arrivals of two process instances

Table 2. Some general purpose key performance indicators. KPI, key performance indicator.

KPI	**Description**
waiting time	time measured from enabling a task to the time when task was actually started
processing time	time measured from the beginning to the end of a single process path
cycle time	sum of time spent on all possible process paths considering the frequencies of the path to be taken in a scenario; cycle time corresponds to the sum of processing and waiting times
process cost	sum of all costs in a process instance
resource utilization	rate of allocated resources during the period that was inspected

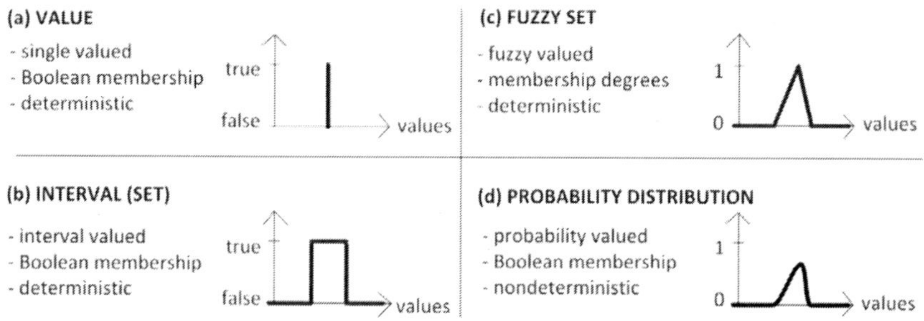

Figure 5. Some different representations of a simulation parameter.

Figure 5c,d represents fuzzy and probabilistic values, respectively. Fuzziness is a type of deterministic uncertainty, which measures the similarity degree of a value to a set. In contrast, probability arises from the question of whether or not a value occurs as a Boolean event. An interval can be considered as a special fuzzy set whose elements have the same similarity to the set. The example of Figure 5c represents a triangular membership function, where there is a unique value that is fully a member of the set (*i.e.*, the abscissa of the triangle vertex). Fuzziness and probability are orthogonal concepts that characterize different aspects of human experience. A detailed theoretical discussion of the relationships between the fuzziness and probability can be found in [40].

An example to show the conceptual difference between probability and fuzzy modeling is described as follows. An insolvent investor is given two stocks. One stock's label says that it has a 0.9 membership in the class of stocks known as non-illegal business. The other stock's label states that it has a 90% probability of being a legal business and a 10% probability of being an illegal business. Which stock would you choose? In the example, the probability-assessed stock is illegal, and the investor is imprisoned. This is quite plausible, since there was a one in 10 chance of it being illegal. In contrast, the fuzzy-assessed stock is an unfair business, which, however, does not cause imprisonment. This also makes sense, since an unfair business would have a 0.9 membership in the class of non-illegal business. The point here is that probability involves conventional set theory and does not allow for an element to be a partial member in a class. Probability is an indicator of the frequency or likelihood that an element is in a class. Fuzzy set theory deals with the similarity of an element to a class. Another distinction is in the idea of observation. Suppose that we examine the contents of the stocks. Note that, after observation, the membership value for the fuzzy-based stock is unchanged while the probability value for the probability-based stock changes from 0.9 to 1 or 0. In conclusion, fuzzy memberships quantify similarities of objects to defined properties, whereas probabilities provide information on expectations over a large number of experiments.

Simulation functionality is provided by many business process modeling tools based on notations, such as EPC (Event-driven Process Chain) or BPMN. These tools offer user interfaces to specify basic simulation parameters, such as arrival rate, task execution time, cost and resource availability. They allow users to run simulation experiments and to extract statistical results, such as average cycle time and total cost. Table 3provides a summary of the main features of some commercial BPS tools [41].

4. INTERVAL-VALUED BUSINESS PROCESS SIMULATION: SPECIFICATION AND DESIGN

4.1. The Output of the Interval-Valued BPS As an Optimization Problem

In this subsection, the basic requirements of our interval-valued BPS model are formally introduced. We employ an implementation of the parameter in terms of interval-valued variable, which is defined as:

$$\ddot{X} = [\underline{x}, \overline{x}] \in \mathbb{IR},\ \underline{x}, \overline{x} \in \mathbb{R},\ \underline{x} \leq \overline{x}$$

where $\mathbb{IR}$ is the set of all closed and bounded intervals in the real line and *x* and *x* are the boundaries of the intervals. An *F*-dimensional array of parameters is then represented by a vector of interval-valued variables as follows:

$$\ddot{\mathbf{X}} = [\ddot{X}_1, \ldots, \ddot{X}_F] \in \mathbb{IR}^F,\ \ddot{X}_i \in \mathbb{IR}$$

Table 3. A summary of commercial business process simulation (BPS) tools supporting BPMN.

BPS Tool Engine	**Short Functional Description**
ARIS (Architecture of Integrated Information Systems) Business Simulator	Locate process weaknesses and bottlenecks; identify best practices in your processes; optimize throughput times and resource utilization; determine resource requirements, utilization levels and costs relating to workflows; analyze potential process risks; establish enterprise-wide benchmarks
Bizagi BPM (Business Process Management) Suite	Identify bottlenecks, over-utilized resources, under-resourced elements in the process and opportunities for improvement. Create multiple what-if scenarios
Bonita Open Solution	Generate business process simulation reports that detail per iteration and cumulative process duration, resource consumption, costs and more

Sparx Systems Enterprise Architect	Identify potential bottlenecks, inefficiencies and other problems in a business process
IBM Business Process Manager	Simulate process performance; identify bottlenecks and other issues; compare actual process performance to simulations; compare simulations to historical performance data; simultaneously analyze multiple processes from a single or multiple process applications
iGrafx Process	Mapping real time simulation to show how transactions progress through the process highlighting bottlenecks, batching and wasted resources. Process improvement and what-if analysis, resource utilization, cycle time, capacity and other pre-defined and customizable statistics
Visual Paradigm Logizian	Simulate the execution of business process for studying the resource consumption (e.g., human resources, devices, *etc.*) throughout a process, identifying bottlenecks, quantifying the differences between improvement options which help study and execute process improvements

MEGA Simulation	Customizable dynamic simulation dashboards; build multiple simulation scenarios for isolated business processes or for covering entire value chains of coordinated business processes; indicators and associated customizable metrics link simulation results with the objectives of the optimization project; indicators can be defined by simple mathematical formulations, such as an MS-Excel formula, or by complex Visual Basic algorithms. Comparisons between indicators and simulation objectives assess the relevance of proposed scenarios
Signavio Process Editor	Visualize process runs and run analysis based on configurable one-case and multiple-case scenarios in order to gain information about cost, cycle times, resources and bottlenecks
TIBCO Business Studio	Perform cost and time analysis and examine workload requirements for the system

It is worth noting that real numbers can be considered as special intervals, called degenerated intervals, with $x = x$ [42].

Let $f : \mathbb{R}^F \rightarrow \mathbb{R}$ be an F-dimensional single-valued business process model. The corresponding interval-valued model $\ddot{f} : \mathbb{IR}^F \rightarrow \mathbb{IR}$ is defined as follows:

$$\ddot{f}(\ddot{\mathbf{X}}) \equiv \ddot{f}(\ddot{X}_1, \ldots, \ddot{X}_F) = \ddot{Y} \equiv [\underline{y}, \overline{y}] \tag{3}$$

where:

$$\begin{aligned} \underline{y} &= \min f(x_1, \ldots, x_F)\big|_{x_i \in \ddot{X}_i,\ \forall i \in \{1,\ldots,F\}} \\ \overline{y} &= \max f(x_1, \ldots, x_F)\big|_{x_i \in \ddot{X}_i,\ \forall i \in \{1,\ldots,F\}} \end{aligned} \tag{4}$$

Formulas (3) and (4) represent the interval-valued BPS as an optimization problem. Figure 6 shows a simple case of interval-valued BPMN model, made of three tasks and two exclusive gateways. Here, each task is measured by an interval-valued parameter [x_i, x_i] (e.g., duration [12,15] min.), as well as the branching proportion of the upper flow of the XOR splitting gateway (e.g., [70,75] %). It is worth noting that the lower flow is constrained by the upper flow, because the total flow must be 100%. For this reason, there is a unique independent interval parameter associated with the gateway, *i.e.*, $\ddot{X}_2$.

For a given set of input interval parameters and a given output KPI (y), the interval-valued output of the process model [y, y] (e.g., total cycle time) can be theoretically determined by simulating the execution of the model for all possible single-valued parameter values $x_1,\ldots,x_4$.

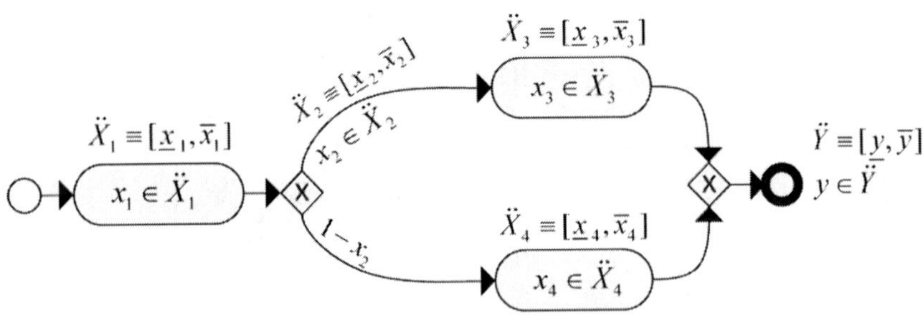

Figure 6. A simple case of interval-valued BPMN model.

Figure 7 gives an exemplification of a mono-dimensional single-valued system function with its interval-valued form, by means of a black solid line and a grey stripe, respectively. Three evaluations of input-output are also included, as single points and rectangles, respectively.

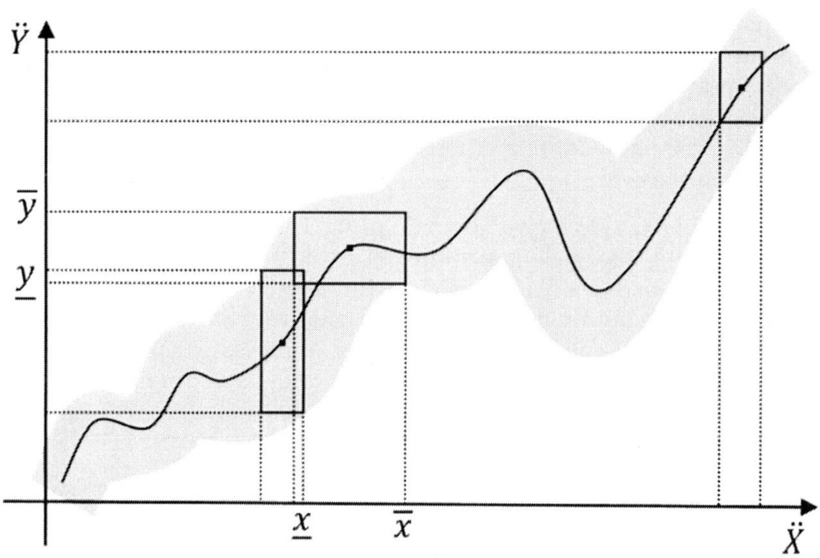

Figure 7. An exemplification of a system with single-valued and interval-valued inputs-outputs.

4.2. The Genetic-Based Optimization

As expressed by Formulas (3) and (4), the problem of evaluating an interval-valued model can be designed as an optimization problem, where the boundaries of the output interval are determined as a minimization and a maximization problem. In our approach, to efficiently compute the interval-valued output of the system, a GA is used. A GA is a search method based on the analogy with the mechanisms of biological evolution [21,43]. A GA is typically made of the following main steps. For a given optimization problem, an initialization process provides a set of randomly generated approximated solutions. Each solution is then evaluated, using a problem-specific measure of fitness. If the termination criteria are satisfied, a solution is then elected as (sub-)optimal for the problem. If not, each solution is encoded as a chromosome (a serialized form). The chromosomes evolve through successive generations (iterations) of the GA. During each generation, a set of new chromosomes, called an offspring, is formed by: (i) the selection of a mating pool, *i.e.*, a quota of parent chromosomes from the current population, according to the fitness values; (ii) a combination of pairs of parents via the crossover genetic operator; and (iii) modification of offspring chromosomes via the mutation genetic operator. The new chromosomes are then decoded in terms of domain solutions. Finally, a new generation is formed by reinserting, according to their fitness, some of the parents and offspring and rejecting the remaining individuals, so as to keep the population size constant [43]. Figure 8 shows the general structure of the GA.

The design of a GA for a given domain problem requires the specification of the following major elements: the chromosome coding of a solution, a fitness function to evaluate a solution and a choice of genetic operators and genetic parameters. The fitness function is represented by the output KPI, the interval-valued output of the process model. The next subsections are devoted to the specification of the other major elements.

4.3. The Genetic Coding

In a GA, a chromosome (or genome) is made of the set of parameters that define a proposed solution to the problem to solve. The chromosome is often represented as a string. Representation is a key issue, because a GA directly manipulates the coded representation of the problem and because the representation schema can severely limit the window by which it observes the search space [44]. Fixed-length and binary coded strings dominate GA research, since there are theoretical results showing their effectiveness and the availability of simple implementations. However, the GA's good properties do not stem from the use of bit strings. One of the most important representations is the real number representation, which is particularly natural to tackle optimization problems with variables in continuous search spaces. More specifically, a chromosome is a vector of floating point numbers whose size is kept the same as the length of the vector, which is the solution to the problem. A GA based on a real number representation is called real-coded GA [44]. A real-coded GA offers a number of advantages in numerical function optimization over binary encodings. The efficiency of the GA is increased, as there is no need to convert chromosomes to phenotypes before each function evaluation; less memory is required as efficient floating-point internal computer representations can be used directly; there is no loss in precision by discretization to binary or other values; and there is greater freedom to use different genetic operators. Figure 9a,b shows the chromosome coding used in the GA. Here, each gene (box) encodes a single-valued parameter of the model. The chromosome contains the minimum number of variables. For instance, a decision point with N outgoing flows is represented via N-1 variables, because a variable can be calculated from the other ones, supposing that their summation must be equal to one.

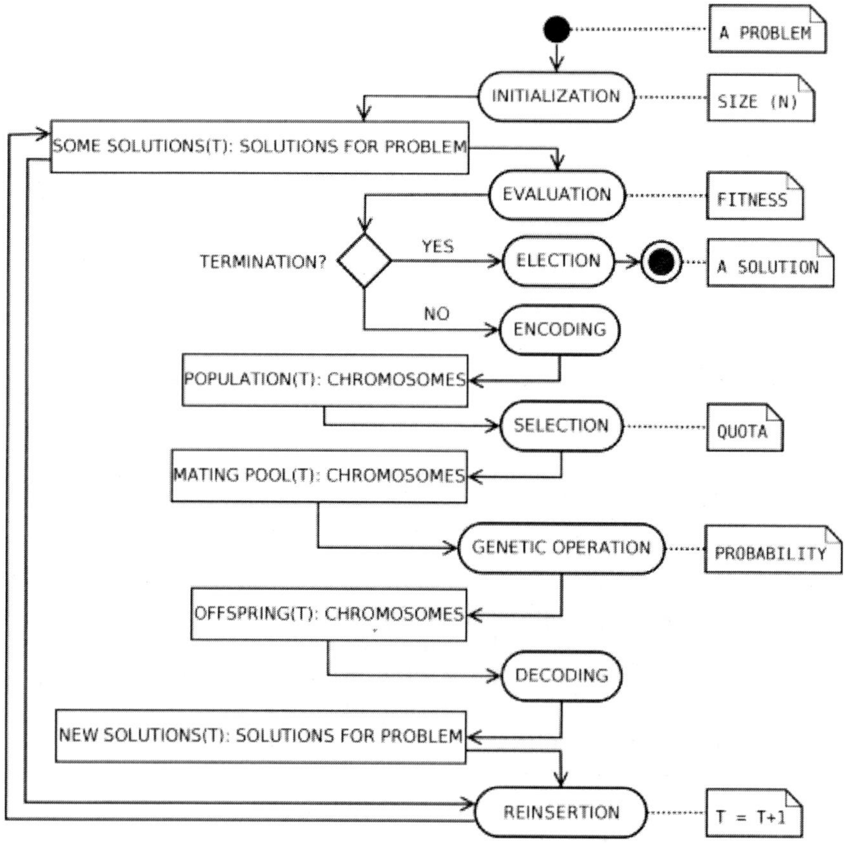

Figure 8. The general structure of the GA.

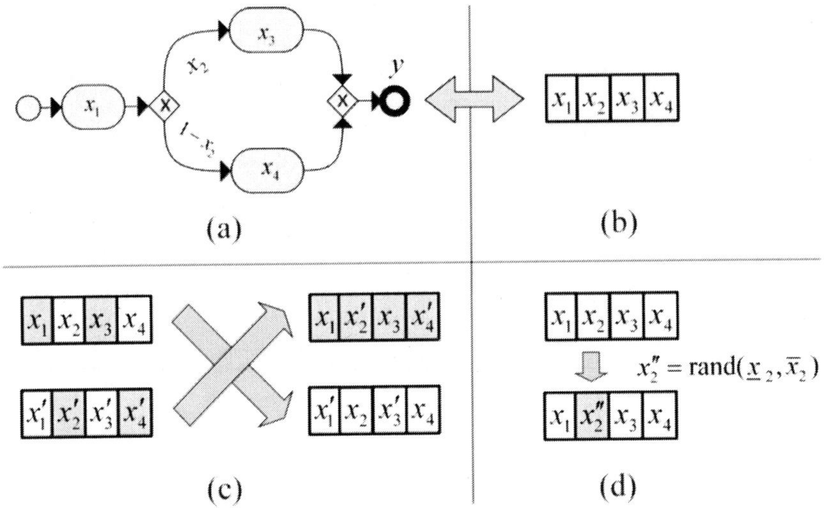

Figure 9. The chromosome coding (**a**, **b**), the crossover (**c**) and the mutation (**d**) operators.

4.4. The Crossover and Mutation Genetic Operators

Crossover is the main genetic operator. It consists of splitting two chromosomes into two or more sub-sequences and obtaining two new chromosomes by exchanging gene sub-sequences between the two original chromosomes. More specifically, we adopted a scattered crossover, in which a binary mask is randomly created of the same length as parents chromosomes. The child is produced by combining parts of chromosomes that lie on the place of Number 1 in the mask from one parent and the rest from the other parent and *vice versa*. Figure 9c shows an example of such a crossover, where zero and one are represented by white and gray colors, respectively. The rationale for this choice is summarized in the following considerations. The idea behind the scattered crossover operator is that the segments contributing to most of the performance of a particular individual may not necessarily be contained in adjacent substrings. Further, the disruptive nature of such a crossover encourages the exploration of the search space, rather than favoring the convergence to highly fit individuals early in the search, thus making the search more robust [21].

The crossover operator is applied with a probability, p_c (crossover rate), on the selected pair of individuals. When the operator is not applied, the offspring is a pair of identical copies, or clones, of the parents. A higher crossover rate allows a better exploration of the space of solutions. However, too high a crossover rate causes unpromising regions of the search space to be explored. Typical values are in the order of 10^{-1} [44].

Mutation is an operator that produces random alteration of the value of a gene, under the constraints enforced by the interval-valued parameter. Figure 9d shows a mutation operator applied on a gene. Mutation is used to maintain genetic diversity. Mutation is randomly applied. The mutation rate, p_m, is defined as the probability that an arbitrary gene is complemented. If it is too low, many genes that would have been useful are never discovered, but if it is too high, there will be much random perturbation, the offspring lose their resemblance to the parents and the GA loses the efficiency in learning from the search history. Typical values of p_m are in the order of 10^{-2} [44]. In our system, we used the adaptive feasible mutation, in which the default mutation function, when there are constraints, randomly generates directions that are adaptive with respect to the last successful or unsuccessful generation. The mutation chooses a direction and step length that satisfies bounds and linear constraints.

4.5. Selection, Reinsertion and Termination Methods

A selection operator chooses a subset of chromosomes from the current population. Various stochastic selection techniques are available. In this work, the roulette wheel method [43] is used. With this method, an individual is selected with a probability that is directly proportional to its fitness. Each individual is mapped to an arc of a circle, whose length equals the individual's fitness. The circumference is then equal to the sum of the fitness. The selection is made by choosing a random number with a uniform distribution between zero and the circumference. The selected individual is the one mapped to the arc containing the chosen point. This ensures that better-fit individuals have a greater probability of being selected; however, all individuals have a chance.

Once a new population has been produced by selection, the crossover and mutation of individuals from the old population, the fitness of the individuals in the new population may be determined. To maintain the size of the original population, the new individuals have to be reinserted into the old population. Similarly, if not all the new individuals are to be used at each generation or if more offspring are generated than the size of the old population, then a reinsertion scheme must be used to determine which individuals are to exist in the new population. If one or more of the best individuals is deterministically reinserted in the next generation, the GA is said to use an elitist strategy. A mutation rate that is too high may lead to the loss of good solutions, unless there is elitist reinsertion. For this reason, we adopted this kind of feature.

As the GA is a stochastic search method, it may be difficult to formally specify convergence criteria. A common practice is to terminate the GA after a prefixed number of generations and then test the quality of the best individual of the population against the problem definition. As the fitness of a population may remain static for a number of generations before a superior individual is found, the application of conventional termination criteria may become problematic. For this reason, we used a more adaptive criterion: the GA terminates if there is no improvement in the best fitness of the population over a predetermined number of generations (called stall generations) or when the preset maximum number of generations is reached.

4.6. Overall Genetic-Based Optimization Algorithm

Let P be the current population, P_M the mating pool (Section 4.2), P_O the offspring (Section 4.2, the set of individuals resulting from crossover and mutation), P_U the set of individuals passed unchanged to the next generation and q and s the cardinality of the mating pool and the cardinality of the current population, respectively. Function select(P) returns an individual from P, selected with the roulette wheel method, and function getPair(P_M) returns a pair of parents from P_M, selected with the roulette wheel method. Function crossover(x, y, p_c) returns the offspring of a pair of parents (Section 4.4). Function mutate(x, P_M) is applied to the selected parent with probability P_M., whereas function reinsert(P, e) returns the first e best individuals from P. Note that the cardinality of the set, P_U, is $s - q$, with individuals drawn from the population, P, passed to the algorithm. It may be observed that all sets used in the algorithm may contain pairs of identical individuals, due to the random character of the various operators. However, each individual is identifiable, even when it is structurally identical to another one. Therefore, all sets are proper sets (not multi-sets). As a consequence, the cardinality of P is a constant.

The optimization algorithm can be formally defined as follows.

Algorithm 1: the genetic-based optimization algorithm.

```
t ← 0;
initialize(P(t));
evaluate(P(t));
while not termination(P(t)) do
    t ← t + 1;
    P(t) = Generation (P(t − 1));
    evaluate(P(t));
end while
return best(P(t));
```

Algorithm 2: a generation of the genetic-based optimization algorithm.

```
Procedure Generation(P)
begin
    P_M ← ∅; P_O ← ∅; P_U ← ∅;
    for i = 1 to q do
        x ← select(P);
        P_M ← P_M ∪ x;
    end for
    for i = 1 to q/2 do
        (x, y) ← getPair(P_M);
        (x', y') ← crossover(x, y, p_c);
        x'' ← mutate(x', p_m);
        y'' ← mutate(y', p_m);
        P_O ← P_O ∪ {x'', y''};
    end for
    for i = 1 to s − q do
        if i ≤ e
            x ← reinsert(P, e);
        else
            x ← select(P, p_c);
        end if
        P_U ← P_U ∪ x;
    end for
    P ← P_O ∪ P_U;
    return P;
end
```

4.7. The Parameter Tuning Process

Some sensitive user parameters for controlling GAs are the following [21,43]: (i) the maximum number of generations allowed; (ii) the population size, which is the number of individuals managed within a generation; (iii) the crossover fraction, *i.e.*, the percentage of chromosomes (from a single generation) that get effectively involved in the crossover operation; and (iv) the elite count, which is the number of individuals with the best fitness values in the current generation that are guaranteed to survive to the next generation. Table 4 summarizes the parameters of the algorithm, together with their typical values. Such parameter values shown have been set according to a generic optimization strategy. In practice, when using evolutionary techniques on a specific case, a number of application constraints narrow down the choice of parameter values. In addition, parameters that cannot be chosen from application constraints can be tuned by using sensitivity analysis. Sensitivity is informally defined as the effect of uncertainty in the parameter on the final results [45].

For the purposes of this paper, application-dependent parameters have been set to the prototypical value of Table 4. With regard to genetic operators, many alternatives are available in the literature. Most of them are used for very specific purposes, unrelated to the aims of this paper.

5. ARCHITECTURAL IMPLEMENTATION

Many GA implementations can be employed in our system. Over the last few decades, considerable research has focused on improving GAs, by producing many types of genetic operators and algorithmic strategies. The interested reader is referred to [21,44] for a comparison of GA methods. It is worth noting that the optimization performed by the IBimp system usually lasts just a few generations. The coexistence of multiple GA implementations is an important requirement in order to allow inspecting the way in which the optimization is carried out, rather than its performance. Indeed, the analyst could be interested in accessing the steps performed by a GA, to better understand the criticalities of some parameters. The management of multiple GAs is, however, intended for advanced users and for experimental purposes, rather than for business users.

Table 4. Parameters of the algorithm and their typical values.

Parameter	Description	Value	Reference
s	population size	50	Section 4.6
p_c	crossover rate	0.7	Section 4.4
p_m	mutation rate	0.1	Section 4.4
e	elite count	2	Section 4.5
q	mating pool size	25	Section 4.6
g_{max}	maximum number of generations allowed	50	Section 4.5
g_{stall}	number of generations considered for stall	20	Section 4.5
ε_{fit}	fitness tolerance to detect an improvement	10^{-6}	Section 4.5

Figure 10 shows how the overall modules of the system are wired together, via a UML component diagram. All the components are executed on the Java Runtime environment. The architecture has been developed and integrated with the MATLAB framework. The source code of our IBimp simulator has been publicly released on the MATLAB Central File Exchange [46] as an extension of Bimp [23] a publicly available java-based business process simulator.

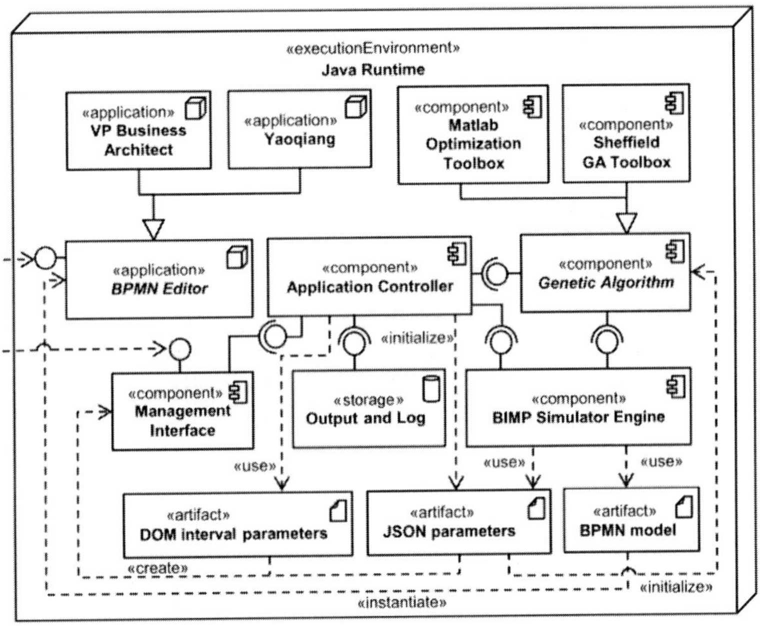

Figure 10. IBimp, overall system architecture.

More specifically, the BP Designer/Analyst (the stick man on the left) first uses a BPMN editor to create a BPMN model (the artifact on the bottom right). For this purpose, any editor can be used, provided that a BPMN interchange format XML file can be exported. We mainly used the Visual Paradigm Business Architect [11] and the Yaoqiang [47] BPMN editors. Once created, the BPMN model artifact can be imported in another software product that supports BPMN interchange XML format. The BPMN model is then enriched by the BP Designer/Analyst with two types of parameters, namely JSON parameters and DOM (Document Object Model) interval parameters, via a management interface. The former represents the additional simulation information of Table 1 and Figure 4, whereas the latter represents the intervals related to some JSON parameters. DOM and JSON are the formats in which such parameters are generated and transferred to other components. More specifically, DOM stands for Document Object Model [48], a cross-platform and language-independent convention for representing and interacting with objects, whereas JSON stands for JavaScript Object Notation, a lightweight data-interchange format [49] that is used to send such parameters to the Bimp Simulator Engine, a core component of the architecture [50]. Bimp is the java-based business process simulator. It takes a file written in BPMN and allows you to add simulation information in JSON format, then outputs process logs and KPIs values. It can be used as a numerical (single-valued) or statistical simulator. We exploited the numerical features only, because the statistical ones are not adequate for our (deterministic) interval-valued approach.

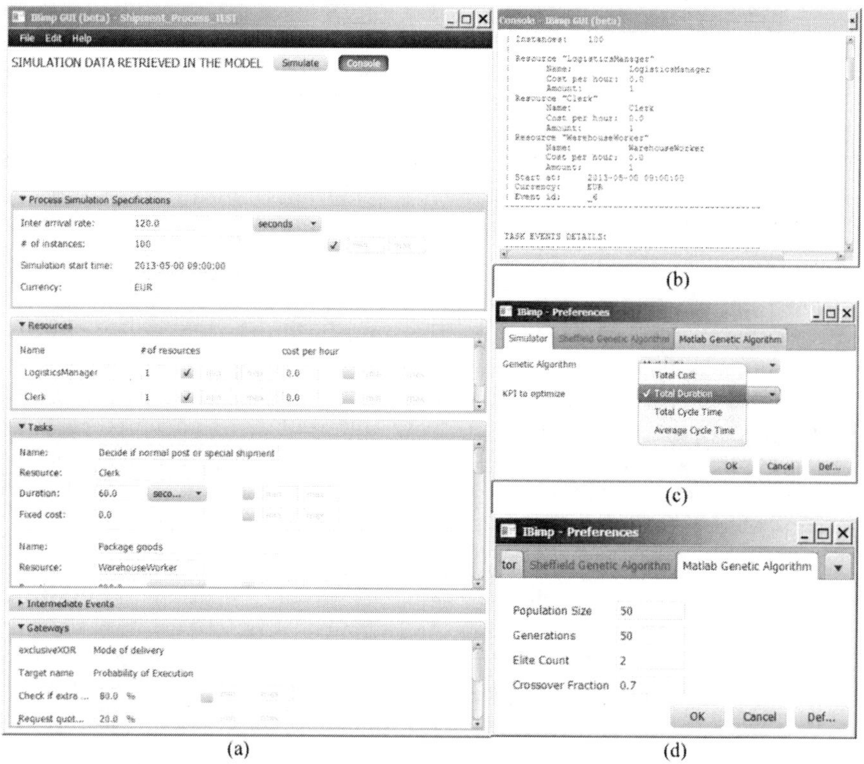

Figure 11. IBimp management interface.

Figure 11 shows the implementation of the management interface. More specifically, Figure 11a shows the main interface for setting both single-valued and interval-valued parameters. Once the XML/BPMN file has been loaded, the main interface shows a number of panes with the input data related to that specific process: general-purpose parameters, resources, tasks, intermediate events and gateways. Further simulation parameters can be set by an advanced user via a plain text configuration file.

Once all parameters have been set, the management interface triggers the Application Controller, a component that acts as an orchestrator, by managing the interconnections and the interactions between the main system components. More specifically, the Application Controller is in charge of: (i) taking the parameters; (ii) executing the core simulation components; and (iii) taking the output and log data. Figure 11b shows the simulation console, with detailed information about the status of the simulation. Figure 11c shows the preferences pane, with the focus on the KPI choice (The Bimp engine currently supports four KPIs: total cost, total duration, total cycle time and average cycle time. In the Java source code of the engine, the class, *KpiCalculator*, can be extended to implement other KPIs). The other output and log data can be derived via log files in comma-separated values (csv) format. Moreover, the system is able to produce some plot of the optimization process, as shown in the experimental section.

The optimization is carried out by the genetic algorithm, a core component that is triggered by the Application Controller. The genetic algorithm is able to control the Bimp Simulator Engine and to instantiate the JSON parameters artifact, so as to carry out a high number of single-valued output evaluations, each with different JSON parameters. When the parameters are all single-valued, the Application Controller does not take into account the genetic algorithm, and the simulation output is computed via the Bimp Simulator Engine only.

The system can be used with different GA implementations. In our implementation, we included the MATLAB Optimization Toolbox [51] and the Sheffield GA Toolbox [52]. Figure 11d shows the common parameters for the MATLAB GA. Further genetic parameters can be set by an advanced user via plain text configuration files. The interested reader is referred to [21,44] for a detailed discussion of the genetic parameters.

6. EXPERIMENTAL STUDIES

Although several works have been published in the field of BPS, unfortunately, there is still a lack of benchmark scenarios in the literature. In this section, we discuss the application of our system to three pilot scenarios and to two real-world scenarios. The aim of this section is to briefly illustrate the possibilities offered by the IBimp simulator rather than to focus on completely solving the scenarios.

In all cases, the simulation is aimed at capturing the process efficiency under a certain variability expressed by some interval-valued parameters. Moreover, each scenario allows exploiting different features of the simulation tool. Depending on the scenario, the total duration and the average cycle time have been used as the main KPIs.

We adopted the MATLAB GA version in the optimization process, by setting the values of population size, elite count and crossover fraction to 50, 2 and 0.7, respectively. We also tested the Sheffield GA version, just to verify that the general properties of the optimization process are kept valid.

6.1. Pilot Scenario: Shipment of a Hardware Retailer

As a first scenario, consider a hardware retailer preparing some steps to fulfill before the ordered goods can actually be shipped to the customer. The scenario comes from [53] and allows testing queuing phenomena in synchronized flows. Figure 12 shows the related BPMN business process diagram. Here, the plain start event "goods to ship" indicates that the preparation is going to be done. After the instantiation of the process, two flows are carried out in parallel: while the clerk has to decide whether this is a normal postal or a special shipment, the warehouse worker can already start packaging the goods. This clerk's task is followed by the exclusive gateway "mode of delivery": only one of the following two branches can be traversed. If a special shipment is needed, the clerk requests quotes from different carriers, then assigns a carrier and prepares the paperwork. If a normal post shipment is fine, the clerk needs to check if an extra insurance is necessary. If that extra insurance is required, the logistics manager has to take out that insurance. In any case, the clerk has to fill in a postal label for the shipment. By means of the inclusive gateway, it can be shown that one branch is always taken, while the other one only if the extra insurance is required. If this is the case, this can happen in parallel to the first branch. Because of this parallelism, we need the synchronizing inclusive gateway right behind "fill in a post label" and "take out extra insurance". Furthermore, we also need the synchronizing parallel gateway before the last task "add paperwork and move package to pick area", to be sure that everything has been fulfilled before the last task is executed.

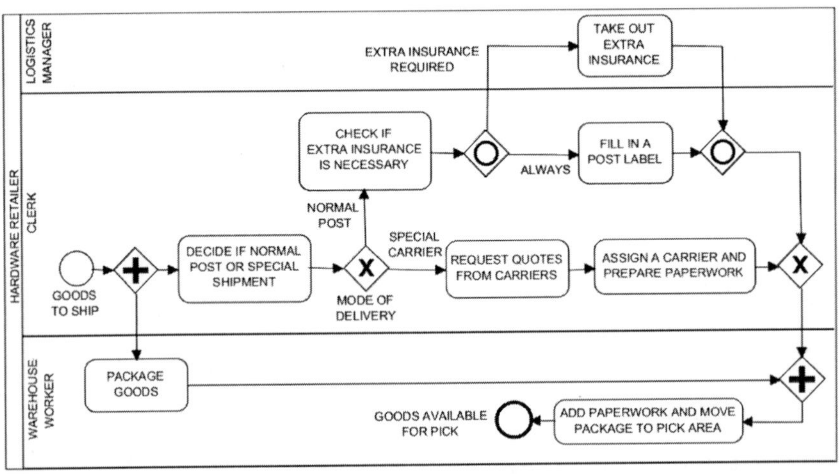

Figure 12. Shipment process of a hardware retailer: BPMN process diagram.

Table 5 shows the model parameters, some of which are interval-valued. First, a single-valued simulation of 100 instances has been carried out, by using single-valued parameters only.

Table 6 and Table 7 show the result. It can be noted that there are very long queues, especially on the last activities, which wait for the management of the extra insurance. Subsequently, an interval-valued simulation has been carried out, by using intervals on some tasks and resources as expressed in the right column of Table 5.

Figure 13a,b shows the optimization processes performed by the GA, which are characterized by a very fast convergence: 15 generations only, carried out in 4 min (results were obtained on a computer with an Intel Dual Core i3-3220, 3.30 GHz, 8GB RAM).

As a result, the total duration interval provided by the system is [577, 4363] h. This outcome is the interval-valued total duration, due to the variability of some parameters in Table 5. Considering the abstraction of Figure 7, the result represent a single evaluation of a specific interval-valued function with 6 input parameters $\ddot{f}_0 : \mathbb{IR}^6 \rightarrow \mathbb{IR}$. By varying the interval-valued parameters on a region of interest, an interval-valued curve similar to Figure 7 can be generated.

Thus, the tool can be used in two ways: (i) as a generator of *characteristic curve* of models subject to uncertainty associated with some input parameters; and (ii) as a tool to evaluate the worst/best cases within a constrained area.

To illustrate the second way of using the system, Table 8 shows the values of the parameters that were used by the GA to have the minimum and maximum KPI, *i.e.*, the best individual of both optimization processes. Here, it can be noted that the KPI minimization occurs by using a high number of resources and a low duration of tasks. Moreover, the KPI maximization occurs by using a low number of resources. This seems logical, since new resources speed up the entire process, whereas tasks duration needs to be adapted to the duration of other synchronized tasks, so as to reduce the queuing phenomena.

Different genetic operators and strategies may find different solutions to the problem. A comparative study can be proficiently handled as future work.

Table 5. Parameters and their values for the shipment scenario.

Parameter Name	Parameter Type	Parameter Value	Parameter Interval
Inter-arrival rate	time rate	120 s	
Decide if normal post	duration	60 s	
Check if extra	duration	60 s	
Request quotes	duration	1300 s	[1200, 2400] s
Assign a carrier &	duration	500 s	[300, 600] s
Take out extra	duration	900 s	[600, 1200] s
Fill in a post label	duration	500 s	
Add paperwork	duration	900 s	
Package goods	duration	900 s	
Normal post	percentage	80%	
Special carrier	percentage	20%	
Extra insurance required	percentage	20%	
Logistics manager	units	1	[1, 5]
Clerk	units	1	[1, 5]
Warehouse worker	units	1	[1, 5]

Table 6. KPIs, by using single-valued parameters, for 100 instances.

KPI	Duration (h)
Shortest execution	7.5
Longest execution	68.7
Total duration	*4289.6*
Average duration	46.9

Table 7. Average waiting time, on 100 instances.

Task	Average Waiting Time (s)
Add paperwork	60,498
Assign a carrier	30,724
Check if extra insurance	28,017
Decide if normal post	9,272
Fill in a post label	31,607
Package goods	39,258
Request quotes	23,799
Take out extra insurance	432

Table 8. Interval-valued results.

Parameter/KPI	Optimization: Final Parameters	
	Minimization	Maximization
Logistics managers	5	5
Clerks	5	1
Warehouse workers	5	1
Request quotes	1605 s	1449 s
Assign a carrier	301 s	557 s
Take out extra	667 s	601 s
Total duration	**577 h**	**4363 h**

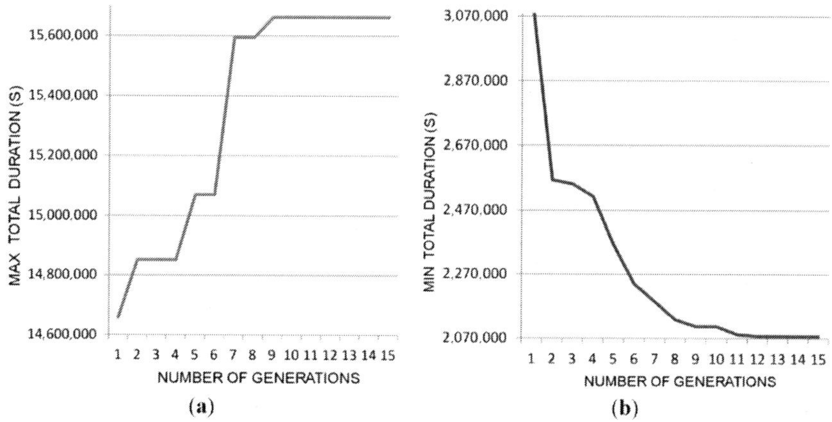

Figure 13. Total duration *vs.* number of generations for the Shipment process.

6.2. Pilot Scenario: Hospital Emergency Center

The second pilot scenario comes from [54] and allows testing models with multiple exclusive flows.Figure 14 shows the BPMN process diagram. Consider the operations of a hospital emergency center (HEC). The process begins when a patient arrives through the admission process in the entrance room of the HEC and ends when a patient is either released from the HEC or admitted into the hospital for further treatment. Patients arriving on their own, after admission, sign in and then are assessed in terms of their condition (Triage). Depending on their condition, patients must then go through the registration process and through the treatment process. Arriving patients are classified into different codes (levels), according to their condition. With the red code, patients are more critical. Patients coming by ambulance are first assigned to the red code and taken to an emergency room (ER) immediately after admission. Once in the room, they undergo their treatment. Finally, they complete the registration process before being either released or admitted into the hospital for further treatment. In contrast, the yellow and green code patients must first sign in with an administrative clerk. After signing in, their condition is assessed by a triage nurse. Such patients must first complete their registration and then go on to receive their treatment. Finally, they are either released or admitted into the hospital for further treatment. The HEC has the following resources: nurses, physicians, technicians, administrative clerks, medical and administrative rooms.

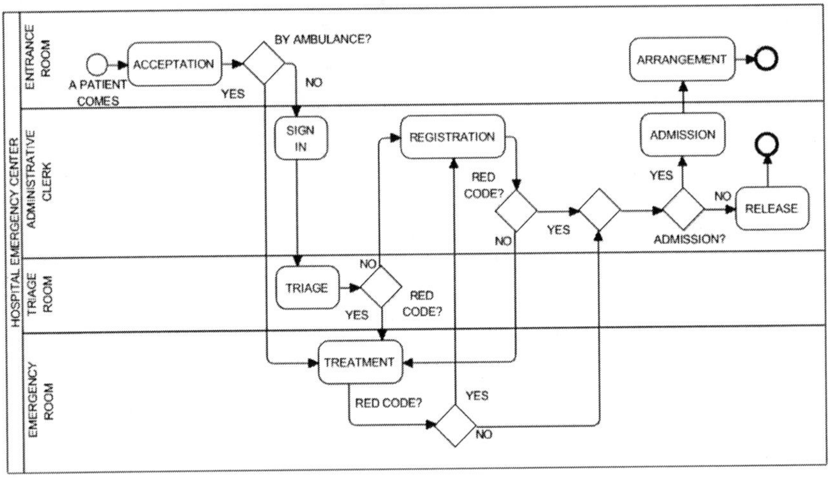

Figure 14. Hospital emergency center: BPMN process diagram.

Table 9 shows the model parameters, five of which are interval-valued. First, a single-valued simulation of 100 instances has been carried out, by using single-valued parameters only. Table 10 shows the result. Subsequently, an interval-valued simulation has been carried out, by using intervals on some tasks and resources, as expressed in the last column of Table 9. Figure 15a,b shows the optimization processes performed by the GA. In terms of convergence, 50 generations were carried out. The total simulation lasted 12 min (results were obtained on a computer with an Intel Dual Core i3-3220, 3.30 GHz, 8GB RAM).

As a result, the total duration interval provided by the system is [62.25, 124.26] h. Table 11 shows the values of the parameters, which were used by the GA to have the minimum and maximum KPI, *i.e.*, the best individual of both optimization processes. It can be noted that the KPI minimization occurs by using a high number of emergency rooms and a low number of triage rooms.

Table 9. Parameters and their values for the hospital scenario.

Parameter Name	Parameter Type	Parameter Value	Parameter Interval
Inter-arrival rate	time rate	900 s	
Acceptation	duration	30 s	
Acceptation	cost	0.24 €	
Sign in	duration	192 s	[100, 200] s
Sign in	cost	1.55 €	
Triage	duration	324 s	[150, 400] s
Triage	cost	11.95 €	
Registration	duration	570 s	[300, 700] s
Registration	cost	4.59 €	
Admission	duration	223 s	
Admission	cost	1.79 €	
Release	duration	132 s	
Release	cost	1.06 €	
Treatment	duration	1554 s	
Treatment	cost	80.69 €	
Arrangement	duration	1275 s	
Arrangement	cost	10.28 €	
By ambulance?	percentage	6% (yes)	
Red code (after triage)?	percentage	10% (yes)	
Red code (after registration)?	percentage	18% (yes)	
Red code (after treatment)?	percentage	14.5% (yes)	
Admission?	percentage	10% (yes)	
Emergency Room	units	3	[2, 5]
Triage Room	units	2	[1, 3]

Table 10. KPIs, by using single-valued parameters, for 100 instances.

KPI	Duration
Shortest execution	20.8 m
Longest execution	256.3 m
Total duration	*116.3 h*
Average duration	69.8 m

Table 11. Interval-valued results.

Resource/KPI	Optimization: Final Parameters	
	Minimization	Maximization
Emergency room	4	2
Triage room	1	1
Registration	300 s	696 s
Triage	165 s	360 s
Sign in	132 s	165 s
Total duration	**62.25 h**	**124.26 h**

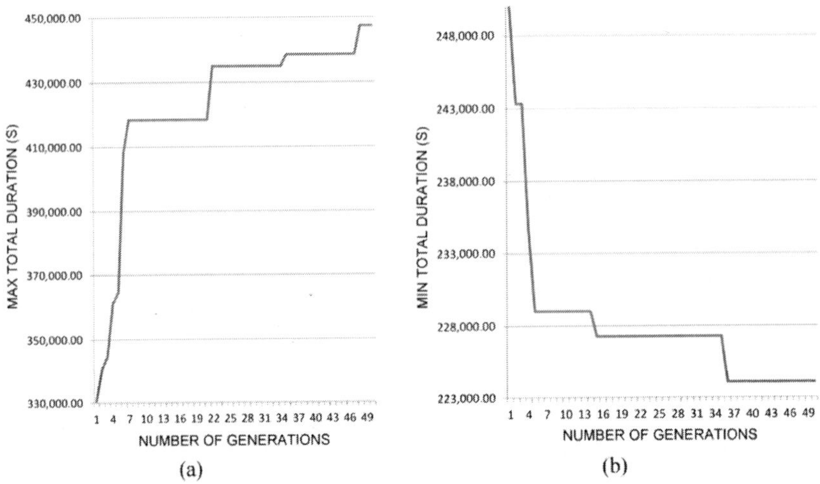

Figure 15. Total duration *vs.* number of generations for the hospital process.

To better investigate the sensitivity of the number of emergency rooms, we can inspect the detailed progress of the KPI during the minimization and maximization steps, shown in Table 12 and Table 13, respectively. More specifically, it is worth noting in Table 12 that a reduction of about 6 h in the total duration is due to a reduction of a handful of seconds of the triage and registration durations, in spite of the fact that the number of triage rooms has also been reduced. It is worth noting in Table 13 that an increase of about 20 h in the total duration can be ascribed to less triage rooms and a small increase of the registration and sign-in durations.

Table 12. KPI along generations during the minimization phase.

Total Duration (h)	Emergency Room	Triage Room	Registration (m)	Triage (m)	Sign in (m)
69.45	4	2	5.51	4.36	1.72
67.59	3	3	5.71	2.75	2.05
67.59	3	3	5.71	2.75	2.05
65.16	3	1	5.71	2.75	2.05
63.61	4	1	5.03	2.75	2.20

Table 13. KPI along generations during the maximization phase.

Total Duration (h)	Emergency Room	Triage Room	Registration (m)	Triage (m)	Sign in (m)
95.76	2	2	11.20	5.98	2.35
100.30	2	2	11.60	5.48	2.74
101.29	2	2	11.60	5.79	2.75
113.32	2	2	11.20	5.98	2.78
116.23	2	1	11.60	5.98	2.75

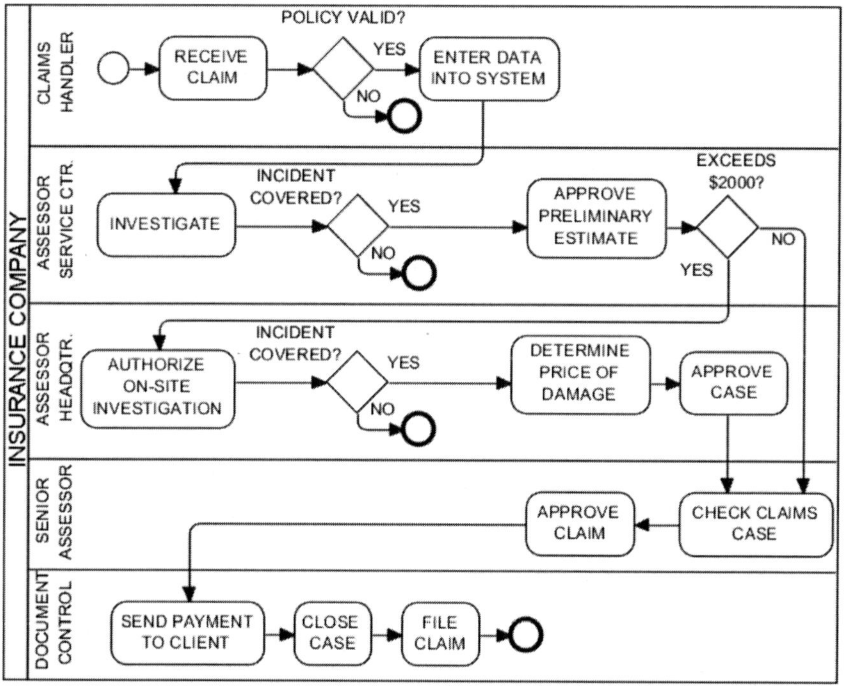

Figure 16. Personal claims process at an insurance company: BPMN process diagram.

6.3. Pilot Scenario: Insurance Company

The third pilot scenario comes from [54] and allows testing optimization in the use of resources. A personal claims department in an insurance company handles claims made by their clients. Figure 16 shows the BPMN process diagram. The first lane corresponds to work done by a claims handler located at the client's local service center. Upon the arrival of a claim, the assessor determines if the client has a valid policy. If not, then the case is terminated; otherwise, the assessor enters the appropriate information in the system. In the second lane, an assessor located at the service center receives data from the claims handler. The assessor first determines if the claim is covered by the client's policy. If not, the case is terminated; otherwise, the assessor approves the preliminary estimate of the damage. If the damage exceeds $2000, the claim is sent to an assessor at headquarters for approval; otherwise, it is sent directly to a senior assessor. Lane 3 corresponds to the assessor at headquarters. The assessor first authorizes the on-site investigation of the accident. If the investigation determines that the incident is not covered by the client's policy, then the case is terminated; otherwise, a final price is determined, and the case is approved. In Lane 4, the senior assessor receives the claim, checks it, completes it and provides the final approval. Once the claim is approved, it is sent to documentation control. Documentation control, in Lane 5, is in charge of processing the payment to the client, closing the case and, finally, filing the claim. The problem in this example is to find the most efficient staffing levels for each of the five resource types.

Table 14 shows the model parameters. First, a single-valued simulation of 100 instances has been carried out. Table 15 shows the resulting KPIs.

Table 14. Parameter and their values for the insurance scenario.

Parameter Name	Parameter Type	Parameter Value
Inter-arrival rate	time rate	5 min
Receive claim	duration	2 min 12 s
Enter data into system	duration	10 min 32 s
Investigate	duration	19 min 28 s
Approve preliminary...	duration	3 min 54 s
Authorize on-site...	duration	2 min 46 s
Determine price of...	duration	37 min 16 s
Approve case	duration	2 min 10 s
Check claims case	duration	3 min 38 s
Approve claim case	duration	1 min 18 s
Send payment to client	duration	7 min 11 s
Close case	duration	1 min 46 s
File claim	duration	3 min 22 s
Policy valid?	percentage	95% (yes)
Incident covered (Assessor Service Center)?	percentage	95% (yes)
Incident covered (Assessor Headquarter)?	percentage	98% (yes)
Exceeds $2000?	percentage	35% (yes)
Claims handler	units	10
Assessor service center	units	10
Assessor HQ	units	10
Senior assessor	units	10
Document control	units	10

Table 15. KPIs, by using single-valued parameters, for 100 instances.

KPI	Duration
Shortest execution	132 s
Longest execution	95.6 m
Total duration	*107 h*
Average duration	64.3 m

Subsequently, an interval-valued simulation has been carried out, by using intervals [5,20] on resources. The GA converged in 40 generations, carried out in 11 min (results were obtained on a computer with an Intel Dual Core i3-3220, 3.30 GHz, 8GB RAM). As a result, the total duration interval provided by the system is [96.5, 109.7] h. Surprisingly, using fewer resources produces a shorter total duration. Table 16 shows the values of the parameters, which were used by the GA to have the minimum and maximum KPI, *i.e.*, the best individual of both optimization processes. Here, it can be noted that the best is to use a very low number of assessor service centers. Moreover, the main difference between minimization and maximization could be ascribed to the number of assessor HQ resources. This can be easily verified by reducing the intervals and restarting the simulator. Thus, our interval-valued simulator allows forming mappings at different levels of detail and, therefore, at different model resolutions.

Table 16. Interval-valued results.

Resource/KPI	Optimization: Final Parameters	
	Minimization	Maximization
Claims handler	14	16
Assessor Service Center	5	8
Assessor HQ	13	6
Senior assessor	10	11
Document control	19	19
Total duration	**96.5 h**	**109.7 h**

6.4. Real-World Scenario: Marine Container Terminal

This scenario presents a real-world experience of the application of our interval-valued simulation in the context of a marine container terminal located in the Port of Leghorn (Italy). A marine container terminal is the place where containers arriving by sea vessels are transferred to inland carriers, such as trucks, trains and *vice versa*. Every marine container terminal performs four basic functions: receiving, storage, staging and loading for both import and export. Receiving involves container arrival at the terminal, either as an import or export, recording its arrival, retrieving relevant logistics data and adding it to the current inventory. Storage is the function of placing the container in a known and recorded location in order to retrieve it when it is needed. Staging is the function of preparing a container to leave the terminal. The containers that are to be exported are identified and organized so as to optimize the loading process. Import containers follow similar processes, although staging is not always performed. An exception is a group of containers leaving the terminal via rail. Finally, the loading function involves placing the correct container on the ship, truck or other mode of transportation.

In this work, the emphasis has been put on internal logistics. More specifically, the aim of the study was to assess the extent to which information technology solutions can improve the overall efficiency of the process. The as-is model is made of 5 lanes, 22 tasks, 9 gateways and 7 types of resources (different types of machinery). Interval parameters were derived by a number of interviews and measurements. A simulation was carried out by considering 20 individuals in the population, 2000 containers and converged in 6 generations, which lasted 16 min (results were obtained on a computer with an Intel Dual Core i3-3220, 3.30 GHz, 8GB RAM). It is worth noting that this simulation time strongly depends on the relatively high number of containers used in a single Bimp simulation, rather than on the number of Bimp simulations requested by the GA. As a result, a minimum total duration of 86.63 h was evaluated. Moreover, a large waiting time was estimated, mostly due to lack of synchronization between resources.

To improve the process, wireless tracking technologies, such as RFID and GPS, were adopted. The to-be model is made of a number of settings able to improve synchronization between machineries. A new simulation was made considering again 20 individuals in the population, 2000 containers and converged in 17 generations, which lasted 38 min (results were obtained on a computer with an Intel Dual Core i3-3220, 3.30 GHz, 8GB RAM). As a result, a minimum total duration of 18.19 h was evaluated, with a better resource utilization, as shown in Figure 17.

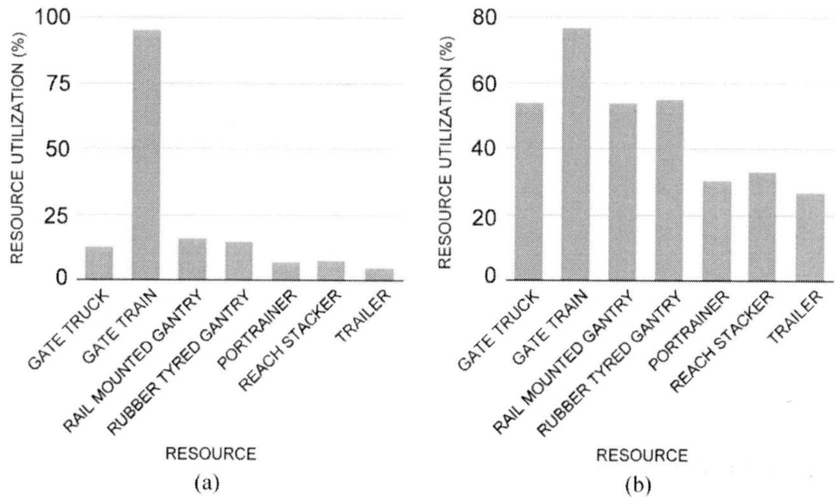

Figure 17. Resource utilization in the as-is (**a**) and to-be (**b**) model.

6.5. Real-World Scenario: Manufacturing Company

This scenario presents a real-world experience of application of our interval-valued simulation tool in the context of the buying process of a leading auto and truck parts manufacturing company.

More specifically, the aim of the study was to improve the overall efficiency of the purchase order process. The as-is model is made of 2 pools, 7 lanes, 24 tasks and 7 gateways. Interval parameters were derived by a number of interviews and measurements. The KPI is the average cycle time (ACT). A single-valued simulation was first performed, considering 25 purchase orders. The resulting ACT was about 42 days. In order to assess this result, an interval simulation was performed, which converged in 15 generations and employed 20 individuals in the population. The simulation lasted 6 min (results were obtained on a computer with an Intel Dual Core i3-3220, 3.30 GHz, 8GB RAM). As a result, the measured ACT was about [28,54] days. Hence, the ACT can be sensibly reduced. Figure 18 shows the parameter values for the as-is, the minimum and maximum KPI.

It can be noted that the maximum KPI may be mainly ascribed to a large duration of the "chief financial officer approval" and "materials and delivery note approval" tasks. This may be easily verified by reducing the intervals and restarting the simulator. Indeed, the interval-valued parameterization allows sensibly reducing the number of trials in incremental investigations.

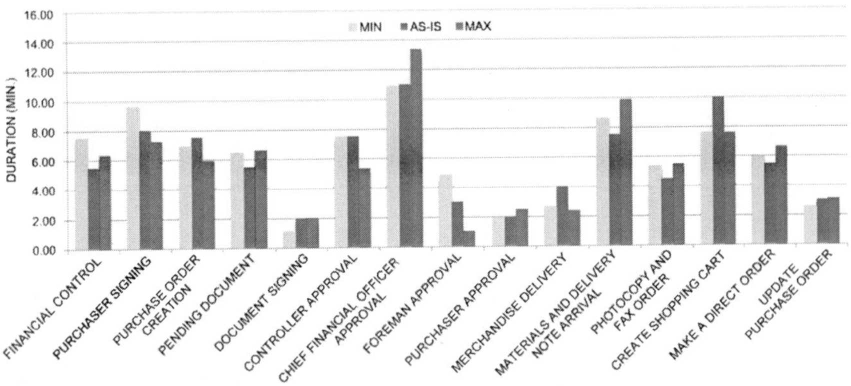

Figure 18. Duration of the relevant tasks, in the minimum KPI, as-is and maximum KPI models.

7. CONCLUSIONS AND FUTURE WORKS

In this paper, a novel approach of business process simulation is presented. The approach is based on the representation of business process parameters in terms of intervals, to efficiently characterize human-driven processes. An interval-valued simulation tool is designed, developed and tested on five scenarios. Given input intervals, the system employs a numerical simulation engine, based on the BPMN standard, and a genetic algorithm to compute the output interval (KPI). The use of BPMN allows the tool to be used by process owners, which is an important condition in BP simulation studies. However, the most important strength point is the possibility of dealing with interval-valued parameters offered by our extension. Indeed, the proposed approach solves a fundamental weakness of existing tools dealing with multi-valued representations, which require an expensive data collection. Moreover, interval-valued data are deterministic and then more suitable for decision-making.

Experimental results have shown that the adopted GA has been an adequate choice, since our approach allowed an efficient analysis of interesting the properties of the observed processes: (i) the worst/best cases within a constrained area; (ii) the parameter values corresponding to the cases of interest; (iii) the importance of some parameter values with respect to the cases of interest; and (iv) the analysis at different levels of detail. Other genetic-based strategies may be effective in solving the type of optimization problem tackled in this work, providing different kinds of solutions. Thus, as future research, we plan to investigate the use of different genetic-based optimization methods, so as to make a comparative analysis. The simulation time performance of the IBimp simulator can be a limitation with respect to the simulation time performance of the lightning fast Bimp simulator, because GAs do not scale well with complexity. In future work, this limitation could be lifted by exploring meta-heuristics, such as simulated annealing, which strike a good trade-off between accuracy and performance. Finally, to conduct performance evaluations of our simulation tool on other KPIs (such as the total cost), as well as on multiple KPIs (*i.e.*, multi-objective optimization) is considered a key investigation activity for future work.

AUTHOR CONTRIBUTIONS

Mario G. C. A. Cimino and Gigliola Vaglini are both responsible for the concept of the paper, the results presented and the writing. Both authors have read and approved the final published manuscript.

CONFLICT OF INTEREST

The authors declare no conflict of interest.

REFERENCES

1. Van der Aalst, W.M.P. Challenges in business process management: Verification of business processes using petri nets. *Bull. Eur. Assoc. Theor. Comput. Sci.* **2003**, *80*, 174–199.

2. Sharp, A.; McDermott, P. *Workflow Modeling*, 2nd ed.; Artech House: Boston, MA, USA, 2009.
3. Barros, A.P., Gal, A., Kindler, E., Eds.; Business Process Management: 10th International Conference, BPM 2012, Tallinn, Estonia, September 3–6, 2012; Proceedings; Lecture Notes in Computer Science. Springer: Berlin/Heidelberg, Germany, 2012; Volume 7481.
4. Object Management Group. Business Process Model and Notation (BPMN), Version 2.0. Official specification. Available online: http://www.omg.org/spec/BPMN/2.0 (accessed on 20 November 2013).
5. Wohed, P.; van der Aalst, W.M.P.; Dumas, M.; ter Hofstede, A.H.M.; Russell, N. On the suitability of BPMN for business process modelling. In *Business Process Management*; Lecture Notes in Computer Sciences; Springer: Berlin/Heidelberg, Germany, 2006; Volume 4102, pp. 161–176.
6. Bozarth, C.C.; Handfield, R.B. *Introduction to Operations and Supply Chain Management*, 3rd ed.; Pearson: Upper Saddle River, NJ, USA, 2013.
7. Anupindi, R.; Chopra, S.; Deshmukh, S.D.; van Mieghem, J.A.; Zemel, E. *Managing Business Process Flows: Principles of Operations Management*, 3rd ed.; Pearson: Upper Saddle River, NJ, USA, 2012.
8. Van der Aalst, W.M.P. A decade of business process management conferences: Personal reflections on a developing discipline. In *Business Process Management*; Lecture Notes in Computer Science; Springer: Berlin/Heidelberg, Germany, 2012; Volume 7481, pp. 1–16.
9. Van der Aalst, W.M.P. Business process simulation revisited. In *Enterprise and Organizational Modeling and Simulation*; Lecture Notes in Business Information Processing; Springer: Berlin/Heidelberg, Germany, 2010; Volume 63, pp. 1–14.
10. Rozinat, A. Process mining. In *Process-oriented Information Systems*; Report on the Seminar; Hasso Plattner Institute for Software Systems Engineering, University of Potsdam: Potsdam, Germany, 2003; pp. 156–171.
11. VP Business Architect. Available online: http://www.visual-paradigm.com/product/lz/ (accessed on 20 November 2013).

12. Bosilj-Vuksic, V.; Hlupic, V. Criteria for the evaluation of business process simulation tools.*Interdiscip. J. Inf. Knowl. Manag.* **2007**, *2*, 73–88.

13. Greasley, A. Using business process simulation within a business process reengineering approach.*Bus. Process Manag. J.* **2003**, *9*, 408–420.

14. Völkner, P.; Werners, B. A simulation-based decision support system for business process planning.*Fuzzy Set. Syst.* **2002**, *125*, 275–287.

15. Van der Aalst, W.M.P.; Nakatumba, J.; Rozinat, A.; Russell, N. Business process simulation: How to get it right? In *Handbook on Business Process Management*; International Handbooks on Information Systems; Springer: Berlin/Heidelberg, Germany, 2010; pp. 317–342.

16. Cimino, M.G.C.A.; Marcelloni, F. Autonomic tracing of production processes with mobile and agent-based computing. *Inf. Sci.* **2011**, *181*, 935–953.

17. Pedrycz, W. *Knowledge-based Clustering: From Data to Information Granules*; Wiley: Hoboken, NJ, USA, 2005.

18. Babuška, R. *Fuzzy Systems, Modeling and Identification*; Technical Report; Delft University of Technology, Department of Electrical Engineering: Delft, The Netherlands, 1999.

19. Hofacker, I.; Vetschera, R. Algorithmical approaches to business process design. *Comput. Oper. Res.***2001**, *28*, 1253–1275.

20. Russell, S.J.; Norvig, P. *Artificial Intelligence: A Modern Approach*, 2nd ed.; Prentice Hall: Upper Saddle River, NJ, USA, 2003; pp. 111–114.

21. Gen, M.; Cheng, R. *Genetic Algorithms and Engineering Design*; Wiley: New York, NY, USA, 1997.

22. Reijers, H.A.; van der Aalst, W.M.P. Short-term simulation: Bridging the gap between operational control and strategic decision making. In Proceedings of the IASTED Conference on Modeling and Simulation, Philadelphia, PA, USA, 5–8 May 1999; pp. 417–421.

23. Bimp-simulator. Available online: http://code.google.com/p /bimp-simulator/ (accessed on 15 May 2014).

24. Wand, Y.; Weber, R. Research commentary: Information systems and conceptual modeling—a research agenda. *Inf. Syst. Res.* **2002**, *13*, 363–376.

25. Van Der Aalst, W.M.P.; ter Hofstede, A.H.M.; Weske, M. Business process management: A survey. In*Business Process Management*; Lecture Notes in Computer Sciences; Springer: Berlin/Heidelberg, Germany, 2003; Volume 2678, pp. 1–12.

26. White, S.A. Introduction to BPMN. Available online: http://www.bptrends.com/bpt/wp-content/ publicationfiles/ 07-04 WP Intro to BPMN - White.pdf (accessed on 20 November 2013).

27. Puhlmann, F. On the suitability of the Pi-Calculus for business process management. In *Technologies for Business Information Systems*; Springer: Berlin/Heidelberg, Germany, 2007; pp. 51–62.

28. Ciaramella, A.; Cimino, M.G.C.A.; Lazzerini, B.; Marcelloni, F. Using BPMN and tracing for rapid business process prototyping environments. In Proceedings of the 11th International Conference on Enterprise Information Systems (ICEIS'09), Milan, Italy, 6–10 May 2009; pp. 206–212.

29. Dumas, M.; La Rosa, M.; Mendling, J.; Reijers, H.A. *Fundamentals of Business Process Management*; Springer: Berlin/Heidelberg, Germany, 2013; pp. 21–23.

30. Vergidis, K.; Tiwari, A.; Majeed, B. Business process analysis and optimization: Beyond reengineering.*IEEE Trans. Syst. Man C. Appl. Rev.* **2008**, *38*, 69–82.

31. Li, L.; Hosking, J.G.; Grundy, J.C. EML: A tree overlay-based visual language for business process modeling. In Proceeding of 9th International Conference on Enterprise Information Systems (ICEIS'07), Funchal, Madeira, Portugal, 12–16 June 2007; pp. 13–19.

32. Van Breugel, F.; Koshkina, M. *Models and Verification of BPEL*; Technical Report. York University: Toronto, ON, Canada, 2006; pp. 1–19. Available online: http://citeseerx.ist.psu.edu/ viewdoc/summary?doi=10.1.1.85.882.

33. Tumay, K. Business process simulation. In Proceedings of the 1996 Winter Simulation Conference (WSC), Coronado, CA, USA, 8–11 December 1996; pp. 93–98.

34. Reijers, H.A. *Design and Control of Workflow Processes*; Springer: Secaucus, NJ, USA, 2003.

35. Wynn, M.T.; Dumas, M.; Fidge, C.J.; ter Hofstede, A.H.M.; van der Aalst, W.M.P. Business process simulation for operational decision support. In Proceedings of the International Workshop on Business Process Intelligence (BPI), Brisbane, Australia, 24 September 2007; pp. 55–66.

36. Pérez-Castillo, R.; Weber, B.; Pinggera, J.; Zugal, S.; García-Rodríguez de Guzmán, I.; Piattini, M. Generating event logs from non-process-aware systems enabling business process mining. *Enterp. Inf. Syst.* **2011**, *5*, 301–335.

37. Van der Aalst, W., Desel, J., Oberewis, A., Eds.; *Business Process Management: Models, Techniques, and Empirical Studies*; Springer: Berlin/Heidelberg, Germany, 2000.

38. Laguna, M.; Marklund, J. *Business Process Modeling, Simulation, and Design*; Pearson Education Dorling Kindersley: New Delhi, India, 2011.

39. Smets, P., Mamdani, E.H., Dubois, D., Prade, H., Eds.; *Non-Standard Logics for Automated Reasoning*; Academic Press: London, UK, 1988.

40. Dubois, D.; Prade, H. Fuzzy sets and probability: Misunderstandings, bridges and gaps. In Proceedings of the 2nd IEEE International Conference on Fuzzy Systems (FUZZ-IEEE'93), San Francisco, CA, USA, 28 March–1st April 1993; Volume 2, pp. 1059–1068.

41. Jansen-vullers, M.H.; Netjes, M. Business process simulation—a tool survey. In Presented at Workshop and Tutorial on Practical Use of Colored Petri Nets and the CPN Tools, Aarhus, Denmark, 24–26 October 2005.

42. Alefeld, G.; Claudio, D. The basic properties of interval arithmetic, its software realizations and some applications. *Comput. Struct.* **1998**, *67*, 3–8.

43. Bernardeschi, C.; Cassano, L.; Cimino, M.G.C.A.; Domenici, A. GABES: A Genetic Algorithm Based Environment for SEU Testing in SRAM-FPGAs. *J. Syst. Architect.* **2013**, *59*, 1243–1254.

44. Herrera, F.; Lozano, M.; Verdegay, J.L. Tackling Real-Coded Genetic Algorithms: Operators and Tools for Behavioural Analysis. *Artif. Intell. Rev.* **1998**, *12*, 265–319.

45. Pinel, F.; Danoy, G.; Bouvry, P. Evolutionary algorithm parameter tuning with sensitivity analysis. In Proceedings of the 2011 International Conference on Security and Intelligent Information Systems (SIIS'11), Warsaw, Poland, 13–14 June 2011; Springer: Berlin/Heidelberg, Germany, 2012; pp. 204–216.

46. Ibimp. Available online: http://www.mathworks.com/matlabcentral/fileexchange/46211-ibimp(accessed on 16 May 2014).

47. Yaoqiang BPMN Editor. Available online: http://bpmn.sourceforge.net (accessed on 20 May 2014).

48. Document Object Model (DOM). Available online: http://www.w3.org/DOM/ (accessed on 20 May 2014).

49. JavaScript Object Notation (JSON). Available online: http://www.json.org (accessed on 20 May 2014).

50. Abel, M. Lightning Fast Business Process Simulator. Master's Thesis, Institute of Computer Science, Faculty of Mathematics and Computer Science, University of Tartu, Tartu, Estonia, 2011.

51. MatLab Optimization Toolbox. Available online: http://www.mathworks.it/it/help/optim/ (accessed on 20 May 2014).

52. Genetic Algorithms Toolbox. Available online: http://codem.group.shef.ac.uk/index.php/ga-toolbox(accessed on 20 May 2014).

53. OMG, Object Management Group. BPMN 2.0 by Example, version 1.0. OMG Document Number dtc/2010–06–02. Available online: http://www.omg.org/spec/BPMN/2.0/examples/PDF (accessed on 20 November 2013).

54. April, J.; Better, M.; Glover, F.; Kelly, J.P.; Laguna, M. Enhancing Business Process Management with simulation optimization. 2005. Available online: http://www.bptrends.com/enhancing-bpm-with-simulation-optimization/ (accessed on 20 November 2013).

7

CHAPTER

OPTIMUM DESIGN FOR CLD LAMINATE PLATES USING GENETIC ALGORITHMS

Guang-Min Luo, Tsung-Yen Hsieh

Department of Naval Architecture and Ocean Engineering, National Kaohsiung Marine University, Kaohsiung, Taiwan

ABSTRACT

The optimizations of constrained layered damped (CLD) laminated structures are discussed in this study. Genetic algorithms (GAs) are employed as the search tool for optimization because these algorithms are suitable for solving optimization problems involving multiple discrete variable combinations. The numerical computation packages, ANSYS and MATLAB, have been used to estimate the optimum stacking sequence of CLD laminated structures. MATLAB package is used to achieve GAs process, and ANSYS package is used to proceed the structural analysis. This study successfully developed a numerical simulation mechanism for optimizing CLD adhesion efficiency by implementing GAs and the finite element method. The loss coefficients of the CLD damping layer vary with vibration frequency and failure constraints of CLD laminated plates are considered in objective function. In addition, the modified plasticity analysis (MPA) is used to increase the search efficiency of GAs and simply plastic analysis.

Keywords: Constrained Layered Damping (CLD), Genetic Algorithms (GAs)

1. INTRODUCTION

When fiber reinforced plastic (FRP) laminate structures are affected by an out-of-plane impact load, delamination frequently occurs due to a lack of interlaminar strength. In recent years, studies regarding how to enhance interlaminar strength have consistently emerged [1] -[3] . In particular, Luo applied constrained layered damping (CLD) to laminates and conducted low-velocity impact experiments and simulations, proving that CLD can effectively enhance the low-velocity impact strength of laminate structures [4] [5] .

CLD is a composite layer made from viscoelastic materials and thin metal. When a structure bends because of external pressure, the viscoelastic layer constrained within the metal layers produces shear deformation, consequently dissipating energy. Theoretically, the larger the amount of deformation is, the more effective CLD is when dissipating energy; however, excessive deformation may induce failure in the structure's base material and the thin metal layers, thereby decreasing structural strength. Therefore, with consideration to the strength of laminate structures, the effectiveness of CLD adhesion can be enhanced by appropriately modifying stacking sequences, facilitating greater deformation capacity of the laminate structure within a reasonable range. This study adopted genetic algorithms (GAs) to examine the optimization of CLD laminate structures, in the hope of developing a numerical method to accurately search for optimal stacking sequences when external impacts occur, and ultimately maximizing the adhesion effects of CLD.

The strength of laminated materials is primarily determined by stacking sequence and angle. In recent years, numerous scholars have developed research on the optimization of laminate structures. A review of relevant studies is as follows:

Kere [6] adopted GAs to simulate the optimization of laminated cylindrical panels under various load conditions; Rao [7] employed scatter search algorithms to calculate the optimal stacking angles of laminates; Matsuzaki [8] adopted the fractal branch-and-bound method to optimize the stacking sequences for asymmetrical laminates; and Aymerich [9] adopted the ant-colony optimization method and identified the maximum value of buckling load by changing the stacking sequences of laminates.

Because stacking sequences and angles of laminated materials are regarded as combinations of multiple discrete variables, performing searches is more difficult when using traditional optimization methods. Hence, a majority of current studies have adopted GAs as a search tool for laminate optimization. For example, after accounting for structural stability, Nagendra [10] employed GAs to identify stacking sequence optimization; Rahul [11] proposed laminate optimization research on lateral impact and combined the finite element method with GAs to design inexpensive lightweight structures; and Antonio [12] contended that composite materials show nonlinear behavior and, consequently, used genetic manipulation to enhance overall search efficiency when optimizing stacking sequences. Moreover, Jafari [13] adopted GAs and finite element analysis to formulate a laminate structure optimization design, subsequently developing a mass-deflection or mass-cost minimization method before analyzing the maximum stiffness of the structures' outer shells. Additionally, Narita [14] employed GAs and targeted staple fiber laminate structures with local anisotropism to conduct vibration optimization, successfully improving the effect of anisotropic staple fibers and achieving optimized designs.

This study adopted CLD adhesion efficiency in laminate structures as an objective for optimized design and searched for the most suitable stacking sequences when specific external forces were applied. This study employed GAs as the search tool for optimization because these algorithms are suitable for solving optimization problems involving multiple discrete variable combinations.

2. METHODS

2.1. Introduction to Analysis

In this study, the ANSYS and MATLAB packages were combined to assess the optimization of CLD laminate structures. Specifically, the GA program was written using MATLAB, whereas structural strength was computed in ANSYS. In addition, the encoding and decoding method adopted in this study differed from those used in traditional GAs: encoding of differing stacking angles was conducted using a ranking method of 0 and 1. The codes and the corresponding stacking angles are shown in Table 1.

The procedures used in the study analysis were as follows: First, random genetic encoding was conducted in MATLAB and the initial population was formed. Next, the codes were converted into their corresponding stacking angles, and structural strength calculations were conducted by inputting parametric design language into ANSYS. After completing structural analysis, data such as the amount of deformation, stress, and strain on the CLD laminate structures were extracted and substituted into MATLAB to compute the GA's objective function and assess fitness values. Simultaneously, a single generation of GA computation was completed. The GA subsequently considered the fitness value outcomes and conducted processes including crossover and mutation, preserving the more optimal angle configurations to breed the next-generation population. Subsequently, the GA repeated computations regarding structural strength and the objective function until no objective functions demonstrating superior optimization levels could be obtained. Finally, the optimal solution was represented as the corresponding stacking angle. The procedures used in this study are shown in Figure 1.

Table 1. GA encoding of stacking angles.

Encoding	Stacking angle	Encoding	Stacking angle
0 0 0	0°	0 1 1	−45°
0 0 1	30°	0 1 0	60°
1 0 0	−30°	1 0 1	−60°
1 1 0	45°	1 1 1	90°

2.2. Objective Function

This study assumed that when a CLD laminate structure is affected by an external impact, as the amount of deformation increases, adhesion becomes more effective. However, common optimization methods determine the minimum value of the objective function when locating optimal solutions. Therefore, when the amount of deformation for the CLD laminate structure reaches Δ, Objective function F can be shown as follows:

$$F = 1 - \Delta \tag{1}$$

Furthermore, to prevent the GA from searching for the optimal solution in a direction of extreme deformation, constraints must be implemented to control the search direction during optimization. When applying GAs to identify the optimal stacking sequence of laminates, constraints cannot be directly established on relevant parameters. Consequently, constraints must be written into the objective function. In this study, the constraints employed were the interlaminar strength of the laminated material, the tensile strength of the CLD, and the damping vibration frequency. The constraining method is described below:

1) Laminated Material Failure Constraint

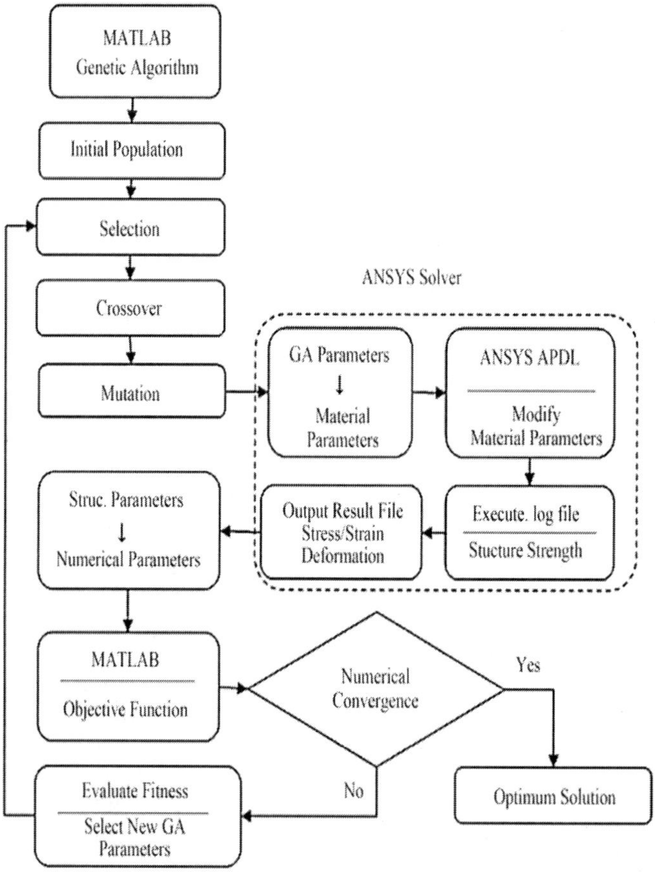

Figure 1. Procedures of optimization.

Delamination is the most common type of failure that results from low-velocity impact. This study referred to the reciprocal stiffness degradation mechanism proposed by Huang [15] and adapted the delamination failure criteria proposed by Yeh, which adopt strain as a parameter, to serve as the failure constraint conditions:

Delamination failure from tension $\varepsilon_{33} > 0$

$$\left(\frac{\varepsilon_{33}}{Z_T^{\varepsilon}}\right)^2 + \left(\frac{\gamma_{13}}{S_{13}^{\varepsilon}}\right)^2 + \left(\frac{\gamma_{23}}{S_{23}^{\varepsilon}}\right)^2 = f \tag{2}$$

Delamination failure from shear $\varepsilon_{33} < 0$

$$\left(\frac{\gamma_{13}}{S_{13}^{\varepsilon}}\right)^2 + \left(\frac{\gamma_{23}}{S_{23}^{\varepsilon}}\right)^2 = f \tag{3}$$

where, 1, 2, and 3 represent fiber direction, perpendicular fiber direction, and thickness direction, respectively. Z_T^{ε} is the strength of tensile strain in the thickness direction, S_{ij}^{ε} is the strength of shear strain on the Plane ij, ε_{ij} is the principal strain in three directions represented on Plane ij, and γ_{ij} is the shear strain on Plane ij.

f is the failure factor and represents the degree of failure. Failure begins when f = 1.0, and f > 1.0 indicates severe failure. The numerical definition is as follows:

$$L_{FC} = \begin{cases} 1.0, & f < 1 \\ 1/f, & f > 1 \end{cases} \tag{4}$$

2) CLD Failure Constraint CLD is composed of damping material and thin metal. As identified in Luo [16] , damping failure occurs subsequent to thin metal failure when a material is affected by impact load. Consequently, when investigating CLD failure, only thin metal failure was considered.

The failure criteria proposed by Von Mises were adopted to form yield judgment criteria for the thin metal:

$$(\sigma_{11}-\sigma_{22})^2+(\sigma_{22}-\sigma_{33})^2+(\sigma_{33}-\sigma_{11})^2=2\sigma^2 \tag{5}$$

σ_{11}, σ_{22}, and σ_{33} represent principal stress in three directions. Assuming that σ_y is the yield stress for pure aluminum, its numerical definition may be defined as follows:

$$C_{FC}=\begin{cases}1.0, & \sigma<\sigma_y\\ \sigma_y/\sigma, & \sigma>\sigma_y\end{cases} \tag{6}$$

3) Vibration Frequency The loss coefficients of the CLD damping layer vary with vibration frequency. This study employed 3M SJ- 2552 as the CLD material, with ISD-112 acting as the viscoelastic glue; the relationship between the vibration frequency of ISD-112 and damping is shown in Figure 2. The natural vibration frequency of the sample employed in this study varied depending on the stacking method, ranging between approximately 400 and 1000 Hz. Therefore, according to Figure 2, the variation in the relationship between Loss Factorη and Frequency ω can be delineated as follows:

$$\eta=0.95-\frac{\omega}{4000} \tag{7}$$

Moreover, the vibration frequency is represented as the following expression after conversion to numerical values:

$$VF = \begin{cases} 1.0, & \eta > \eta_{max} \\ \eta/\eta_{max}, & \text{Otherwise} \end{cases} \tag{8}$$

where, the η_{max} of 3M SJ-2552 is 0.9.

After establishing the quantified constraints, these constraints can then be employed to modify the initial objective function and to further accommodate the GA in its search for the optimal solution. The modified objective function was

$$F = \frac{1-\Delta}{L_{FC} \times C_{FC} \times VF} \tag{9}$$

2.3. Modified Plasticity Analysis (MPA)

This study considered eight lamination angles, 0°, ±30°, ±45°, ±60°, and 90°, to analyze laminar models with 16 layers and dimensions measuring 0.10 m long and 0.05 m wide. The laminated material was Toray T700/Epoxy E765, and the damping material for the constraint layer was 3M SJ-2552. Properties for related materials are shown in Table 2 and Table 3. During finite element simulation, the solid element, Solid 46, was used to simulate the laminar structure, whereas Solid 45 was employed to simulate the damping material and aluminum sheet. Boundary conditions were fixed at both ends.

ANSYS typically conducts simplified plastic analysis and frequently assumes that the elastic-plastic behavior of a material implies a bilinear stress-strain relationship; however, computation using ANSYS is highly timeconsuming. To increase the search efficiency of the GA, this study modified aluminum plasticity stress in MATLAB rather than performing this procedure in ANSYS, and titled the process "modified plasticity analysis" (MPA). The analysis method is described below:

The bilinear stress-strain relationship of the aluminum yield was first written into MATLAB, and ANSYS was implemented to perform only linear elasticity analysis. The bilinear stress-strain relationship of the aluminum yield is shown in Figure 3.

Table 2. Material properties of Toray T700/Epoxy E765.

	Parameters		Parameters
E_{1t}	$129 \times 10^9 N/m^2$	Tensile strength F_{1t} (ASTM D-3039-95)	$2.55 \times 10^9 N/m^2$
E_{1c}	$124 \times 10^9 N/m^2$	Compression strength F_{1c} (ASTM D-3039-95)	$1.24 \times 10^9 N/m^2$
$E_{2t} = E_{3t}$	$9.37 \times 10^9 N/m^2$	Tensile strength $F_{2t} = F_{3t}$ (ASTM D-3039-95)	$0.042 \times 10^9 N/m^2$
$E_{2c} = E_{3c}$	$12.53 \times 10^9 N/m^2$	Compression strength $F_{2c} = F_{3c}$ (ASTM D-3039-95)	$0.185 \times 10^9 N/m^2$
G_{12}	$4.47 \times 10^9 N/m^2$	Shear strength S_{12} (ASTM D-2344)	$0.137 \times 10^9 N/m^2$
$\nu_1 = \nu_2 = \nu_3$	0.335	ρ	$1560 kg/m^3$

Table 3. Material properties of 3M SJ-2552.

	Damper	**Aluminum**
E	5.51×10^{7} N/m^2	6.8×10^{10} N/m^2
ν	0.45	0.36
ρ	980 kg/m^3	2699 kg/m^3
Thickness (ASTM D-3652)	0.13 mm	0.25
Tensile strength (ASTM D-3759)	1400 N/m^2	--
Elongationat break (ASTM D3759)	15%	--

In Figure 3, Point A shows the yield stress of the material, m1 is the coefficient of elasticity for aluminum, and m^2 is the coefficient of elasticity beyond the aluminum yield point. The aluminum stress value was retrieved during each generational computation to determine if it had exceeded the yield stress point. If the aluminum stress had not yet reached the yield stress point, no modification was required. If the aluminum stress had exceeded the yield stress point, it was automatically converted into a stress value corresponding to the m2 curve and substituted into Equations (5) and (6) to solve for parameter C_{FC}. If MPA was proven feasible, it could reduce the time required for performing plasticity analysis using ANSYS as well as significantly improve computational efficiency.

The testing case of MPA was titled "M-Plas-01" and normal plasticity analysis of ANSYS was titled "Plas-01". Subsequently, the results of both methods were compared to verify the accuracy of MPA.

In Table 4, computational results derived from comparisons of Plas-01 and M-Plas-01 regarding factors such as the amount of deformation in the CLD laminar structure and aluminum stress suggest an error value of less than 5%. This demonstrated that M-Plas-01 possessed a significant level of precision during analysis. Furthermore, in the search for optimal solutions, the computation time required for M-Plas-01 was significantly shorter than that necessary for Plas-01, facilitating the completion of computations in half the time required for traditional methods. Consequently, M-Plas-01 demonstrated higher efficiency.

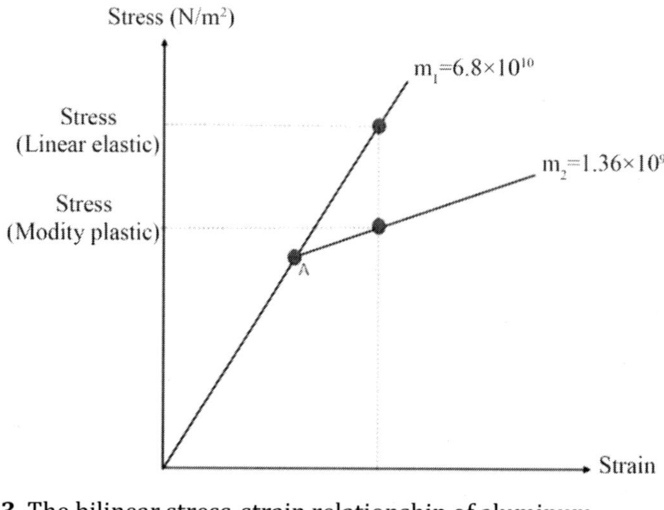

Figure 3. The bilinear stress-strain relationship of aluminum.

Table 4. Analysis results of Plas-01 and M-Plas-01.

Nomenclature	*Plas*-01	*M-Plas*-01
Stacking sequence	[0°/0°/0°/–30°/–30°/45°/30°/90° /0°/0°/0°/–30°/0°/0°/30°/45°]	[30°/0°/0°/0°/–30°/30°/–45/–45° /0°/0°/0°/45°/0°/–30°/–30°/45°]
Deformation	0.891 mm	0.884 mm
Stress of Al.	40.9 MPa	40.8 MPa
Objective function	1.0227	1.0193
Time required	614.37 mins	333.18 mins

Based on these test results, traditional plasticity analysis was replaced with the proposed modified plasticity analysis in subsequent optimization computations to further reduce computation time.

3. OPTIMIZATION PARAMETERS

When assigning parameters for optimization, varying genetic parameters may affect the convergence and accuracy of numerical analysis. Therefore, a uniform static load of 105 N/m^2 was initially employed to determine appropriate genetic parameters, which then served as a reference when identifying the optimal solution for dynamic load.

The parameters considered in this study included population size, number of genes, crossover rate, mutation rate, and elitism. Subsequently, each parameter was adjusted and modified to facilitate comparisons of the computation results and the time required to complete computations.

The order of parameter modifications was as follows: no crossover rate (Cros-off), high crossover rate (Croshigh), no mutation rate (Mut-off), high mutation rate (Mut-high), small population size (Popu-less), large population size (Popu-high), small generation size (Gen-less), large generation size (Gen-high), and no elitism (Eli-off). The assignment of parameter values is shown in Table 5.

The computation results for the parameters are shown in Table 6. Comparisons between the target functions and time demonstrated that Cros-high, Mut-off, and Mut-high possessed significant search efficiency. However, parameter assignment for Mut-off was not adopted because Mut-off implied that no mutation rate was set, and an assumption of no mutation rate could result in a partial optimal solution. Furthermore, optimal solutions can be obtained when fewer generations of Mut-high are employed; consequently, the Mut-high assigning mode was adopted in subsequent dynamic simulations to increase efficiency when identifying optimal solutions.

Table 5. Assignment of gene parameter values.

	Generation	Population	Gene number	Crossover rate	Mutation rate	Elitism
Cross-off	350	8	48	0	0.05	○
Cross-high	350	8	48	1	0.05	○
Mut-off	350	8	48	0.8	0	○
Mut-high	350	8	48	0.8	0.1	○
Popu-less	350	4	48	0.8	0.05	○
Popu-high	350	16	48	0.8	0.05	○
Gen-less	75	8	48	0.8	0.05	○
Gen-high	450	8	48	0.8	0.05	○
Eli-off	350	8	48	0.8	0.05	×

Table 6. Computation results for disparate gene parameters.

	Deformation (mm)	Objective function	Computing time (min)	Convergence generation
Cross-off	1.0	1.0267	311.67	167
Cross-high	0.809	1.0513	334.81	291
Mut-off	0.820	1.0166	300.12	349
Mut-high	0.846	1.0174	327.34	262
Popu-less	0.942	1.0220	262.12	332
Popu-high	0.782	1.0148	610.93	307
Gen-less	0.948	1.0214	71.81	54
Gen-high	0.812	1.0164	383.28	417
Eli-off	1.1	1.0289	328.55	317

4. DISCUSSIONS

This study primarily endeavored to investigate the structural optimization of CLD laminar structure when affected by low-velocity impact. The low-velocity impact conditions were simulated using dynamic uniform loadings and assuming that impact duration (Δ) of uniform loadings were 0.01 s and 0.02 s individually. Additionally, uniform loadings were postulated to impact at a half sine-wave, with maximum values of impact force (P_{max}) were 500 N and 1000 N. Types of external force are shown in Figure 4.

Structural strength and stiffness can be determined using the stacking sequence of a laminate. Therefore, when the deformation of a laminate increases or when the laminate is located in a certain vibration frequency range, the damping layer can demonstrate larger shear deformation or a higher loss factor, both of which would affect CLD adhesion efficiency. Thus, the optimization objective for this study was to adequately modify stacking angles so that an appropriate amount of deformation in the laminate was produced when it was affected by an equivalent load.

During simulation, modal analysis was initially conducted to identify the natural vibration frequency of the CLD laminar structure, thereby allowing the computation of parameter VF in the objective function. A dynamic impact simulation was subsequently conducted and the maximum values of the amount of deformation, stress, and strain for the CLD laminar structure were computed; these values varied with time. The values of the calculated stress and strain could be used to compute the L_{FC} and C_{FC} values in the objective function, and the amount of deformation was represented as Δ in the objective function.

Subsequently, the damping thickness was modified and adopted as a genetic variable. Damping thickness was entered into the GA with the stacking angle variable to assess and observe if adhesion efficiency improved after modifying both factors. In the present study, the dynamic analysis of constant damping thickness was titled "Opt-n" and varying damping thickness was titled "Opt-n-d". The code n was 1 to 4, and it represented various impact loading conditions respectively. The assignment of genetic parameters for Opt-n-d was identical to that for Opt-n, in which damping thickness was constrained between 0.1 mm and 1.0 mm. After computation, the parameters of Opt-n and Opt-n-d, such as the objective function and the amount of deformation, were compared, and this comparison was adopted to assess whether significant variation occurred in adhesion efficiency due to modifications in damping thickness. Nomenclature of different loading conditions was shown in Table 7.

The computation results for low-velocity impact optimization are shown in Tables 7-10. In this study, Opt-1 and Opt-1-d are utilized to illustrate the computation results first, and the search and retrieval processes employed for the GA are shown in Figure 5.

Table 8 shows the computation results of comparisons between Opt-1 and Opt-1-d. A discussion regarding the Opt-1 and Opt-1-d results derived from the objective function and relevant deformation amounts is included below:

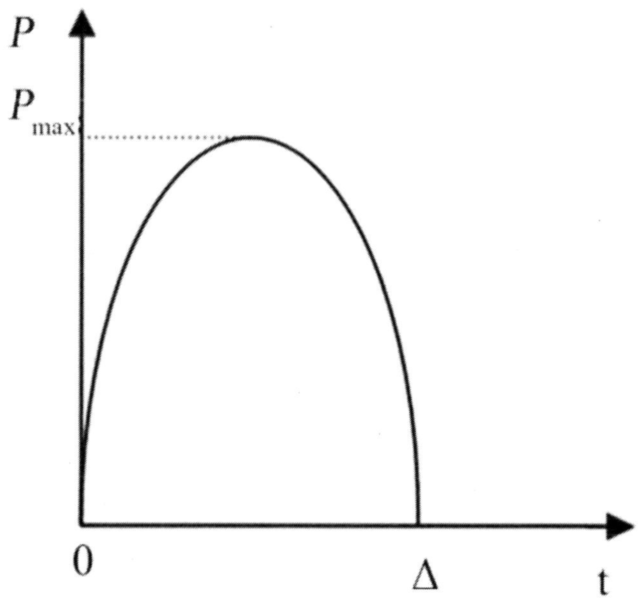

Figure 4. Half-sine wave loading.

Table 7. Nomenclature of various impact loading conditions.

Nomenclature	*Opt*-1/*Opt*-1-d	*Opt*-2/*Opt*-2-d	*Opt*-3/*Opt*-3-d	*Opt*-4/*Opt*-4-d
Loading P_{max}	500 N	500 N	1000 N	1000 N
Duration Δ	0.01 s	0.02 s	0.01 s	0.02 s

*, "-d" means damping thickness is a varying parameter.

Table 8. Comparison results of Opt-1 and Opt-1-d.

Nomenclature	*Opt*-1	*Opt*-1-d
Stacking angle	[−60°/−45°/60°/−60°/60°/−45°/60°/−60° /0°/0°/−45°/0°/0°/30°/−60°/45°]	[45°/90°/60°/45°/60°/30°/30°/45° /0°/0°/30°/30°/−45°/60°/−45°/−45°]
Objective function	1.2581	1.2472
Deformation	3.1 mm	2.6 mm
Damping thickness	0.127 mm	0.882 mm
Natural Frequency	660.1	661.9

Table 9. Comparison results of Opt-2 and Opt-2-d.

Nomenclature	*Opt*-2	*Opt*-2-d
Stacking angle	[45°/−60°/−60°/60°/45°/60°/−60°/60° /0°/0°/0°/30°/−30°/90°/−60°/45°]	[90°/90°/−45°/90°/90°/60°/−60°/ −60°/0°/0°/0°/45°/−60°/45°/60°/45°]
Objective function	1.2454	1.2376
Deformation	2.7 mm	2.6 mm
Damping thickness	0.127 mm	0.868 mm
Natural Frequency	667.9	648.1

Table 10. Comparison results of Opt-3 and Opt-3-d.

Nomenclature	*Opt*-3	*Opt*-3-d
Stacking angle	[0°/−30°/60°/−60°/60°/−45°/−60°/90° /0°/0°/0°/0°/−30°/90°/−60°/60°]	[0°/60°/−30°/90°60°/−45°/45°/−60° /0°/0°/0°/0°/60°/45°/90°/−60°]
Objective function	1.3794	1.3597
Deformation	4.0 mm	3.6 mm
Damping thickness	0.127 mm	0.874 mm
Natural Frequency	655.8	792.4

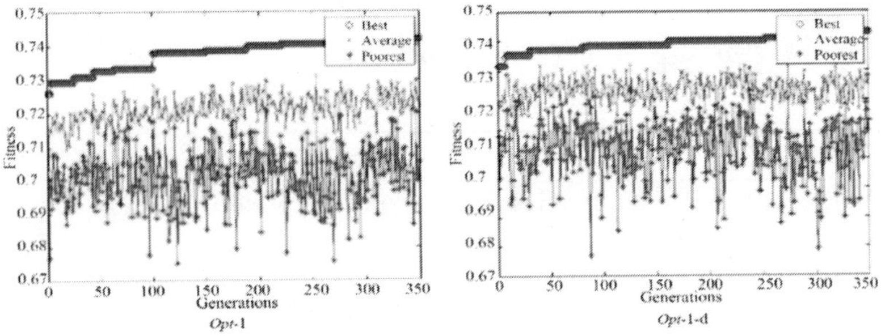

Figure 5. The retrieval processes of Opt-1/Opt-1-d.

First, as the objective function decreases, the adhesion efficiency identified during the GA search increases. Table 7 indicates that the objective functions of Opt-1-d, 1.2472, is slightly smaller than that of Opt-1, thereby proving that modifying damping thickness is effective in improving CLD adhesion efficiency.

In addition, computations for Opt-1-d suggested that only 2.6 mm of deformation was required to achieve optimal CLD adhesion efficiency. This amount was 16.13% less than that computed using Opt-1, which was equivalent to 3.1 mm. This phenomenon confirms that increasing the amount of deformation is not the only method that can be adopted to improve CLD adhesion efficiency and that results of more substantial significance are observed by appropriately increasing damping thickness. However, the computation results also indicate that increasing damping thickness is not necessarily a superior method for optimization, and optimal effects were only produced when the increase in damping thickness was complemented by stacking sequences for the laminated material. This data confirms the necessity of structural optimization.

Finally, regarding damping thickness, the comparison in Table 8 demonstrates that the thickness determined using Opt-1-d was 0.882 mm, which was much larger than the damping layer thickness of 0.127 mm for 3 M SJ-2552. However, the density of the damping material was one-third that of the laminated material. Based on the density of typical laminar structures, this degree of variation in damping thickness should not significantly affect the mass of the overall structure. As indicated by Table 8, appropriately increasing the thickness of the damping layer can reduce structural deformation amounts from 3.1 mm to 2.6 mm under the preliminary condition that similar damping adhesion efficiencies are present. This verifies that increasing damping thickness can maintain optimal adhesion efficiency as well as effectively reduce the amount of deformation in the laminar structure, thereby decreasing the probability of structural failure. Furthermore, the difference in vibration frequency was less than 1%, suggesting that this type of damping possessed a higher energy absorption capacity within the proposed vibration frequency range.

The other three kinds of computation results have the same conclusion as Opt-1 and Opt-1-d presented. The details were shown in Tables 9-11.

Based on the above results, it is evident that this study has successfully employed GAs to evaluate an appropriate stacking angle arrangement, resulting in optimal efficiency when CLD is adhered to laminates.

5. CONCLUSIONS

This study successfully developed a numerical simulation mechanism for optimizing CLD adhesion efficiency by implementing GAs and the finite element method. In addition, appropriately modifying damping thickness had evident benefits concerning enhancements in adhesion efficiency. The methods implemented and conclusions drawn from the simulation analyses conducted in this study are summarized below:

1) MATLAB and ANSYS were successfully integrated to examine the optimization of CLD adhesion efficiency when affected by impacts resulting from specific external forces. Future research can adapt the modal investigation employed in this study to explore the optimal effects of CLD adhesion and the optimal stacking angle arrangement when affected by different external force loads.
2) The modified plasticity analysis (MPA) method, in which stress modification beyond the aluminum yield point is written directly into MATLAB, was proposed. Consequently, finite element analysis required only linear elasticity computations, which effectively reduced the time traditionally required to conduct plasticity analysis using ANSYS and substantially improved speed when searching for optimal solutions.
3) This study confirmed that appropriately modifying both stacking angles and damping thicknesses are beneficial to improve the efficiency of CLD adhesion to laminates. In particular, increasing damping thickness can result in optimal adhesion efficiency in the case of a smaller amount of deformation, thereby enhancing structural stability.

Table 11. Comparison results of Opt-4 and Opt-4-d.

Nomenclature	***Opt*-4**	***Opt*-4-d**
Stacking angle	[30°/−30°/−60°/30°/−30°/−30°/60°/−45° /0°/0°/0°/0°/90°/−30°/45°/45°]	[−30°/30°/60°/−45°/30°/−60°/90°/0° /90°/0°/0°/0°/−45°/−45°/−45°/30°]
Objective function	1.3792	1.3264
Deformation	3.3 mm	2.9 mm
Damping thickness	0.127 mm	0.822 mm
Natural Frequency	794.9	797.9

ACKNOWLEDGEMENTS

The authors would like to thank the National Science Council of the Republic of China for financially supporting this research.

REFERENCES

1. Mujika, F., Vargas, G., Ibarretxe, J., Gracia, J.D. and Arrese, A. (2012) Influence of the Modification with MWCNT on the Interlaminar Fracture Properties of Long Carbon Fiber Composites. Composites Part B, 43, 1336-1340.
2. Kousourakis, A., Mouritz, A.P. and Bannister, M.K. (2006) Interlaminar Properties of Polymer Laminates Containing Internal Sensor Cavities. Composite Structures, 75, 610-618.
3. Li, Y., Hori, N., Arai, M., Hu, N., Liu, Y. and Fukunaga, H. (2009) Improvement of Interlaminar Mechanical Properties of CFRP Laminates Using VGCF. Composites Part A, 40, 12, 2004-2012.
4. Luo, G.M. and Lee, Y.J. (2011) Quasi-Static Simulation of Constrained Layered Damped Laminated Curvature Shells Subjected to Low-Velocity Impact. Composites Part B, 42, 1233-1243.
5. Luo, G.M. (2012) Estimate of CLD Increase Laminated Plate and Shell Low-Velocity Impact Strength. World Journal of Engineering, 9, 319-330.
6. Kere, P. and Koski, J. (2001) Multicriterion Stacking Sequence Optimization Scheme for Composite Laminates Subjected to Multiple Loading Conditions. Composite Structures, 54, 225-229.
7. Rama Mohan Rao, A. and Arvind N. (2005) A Scatter Search Algorithm for Stacking Sequence Optimization of Laminate Composites. Composite Structures, 70, 383-402.
8. Matsuzaki, R. and Todoroki, A. (2007) Stacking-Sequence Optimization Using Fractal Branch-and-Bound Method for Unsymmetrical Laminates. Composite Structures, 78, 537-550.
9. Aymerich, F. and Serra, M. (2008) Optimization of Laminate Stacking Sequence for Maximum Buckling Load Using the Ant

Colony Optimization (ACO) Metaheuristic. Composites: Part A, 39, 262-272.

10. Nagendra, S., Haftka, R.T. and Gurdal, Z. (1992) Stacking Sequence Optimization of Simply Supported Laminates with Stability and Strain Constraints. AIAA Journal, 30, 2132-2137.

11. Rahul, Sandeep, G., Chakraborty, D. and Dutta, A. (2006) Multi-Objective Optimization of Hybrid Laminates Subjected to Transverse Impact. Composite Structures, 73, 360-369.

12. António, C. (2006) A Hierarchical Genetic Algorithm with Age Structure for Multimodal Optimal Design of Hybrid Composites. Structure Multidisc Optimization, 31, 280-294.

13. Azarafza, R., Khalili, S.M.R., Jafari, A.A. and Davarb, A. (2009) Analysis and Optimization of Laminated Composite Circular Cylindrical Shell Subjected to Compressive Axial and Transverse Transient Dynamic Loads. Thin-Walled Structures, 47, 970-983.

14. Honda, S. and Narita, Y. (2011) Vibration Design of Laminated Fibrous Composite Plates with Local Anisotropy Induced by Short Fibers and Curvilinear Fibers. Composite Structures, 93, 902-910.

15. Lee, Y.J. and Huang, C.H. (2003) Ultimate Strength and Failure Process of Composite Laminated Plates Subjected to Low-Velocity Impact. Reinforced Plastics and Composite, 22, 1059-1081.

16. Luo, G.M. and Lee, Y.J. (2009) Simulation of Constrained Layered Damped Laminated Plates Subjected to Low-Velocity Impact Using a Quasi-Static Method. Composite Structures, 88, 290-295.

8

CHAPTER

DEVELOPMENT AND COMPARISON OF HYBRID GENETIC ALGORITHMS FOR NETWORK DESIGN PROBLEM IN CLOSED LOOP SUPPLY CHAIN

Muthusamy Aravendan[1], Ramasamy Panneerselvam[2]

[1]*Department of Leather Design (Footwear & Products), National Institute of Fashion Technology, Chennai, India*

[2]*Department of Management Studies, School of Management, Pondicherry University, Pondicherry, India*

ABSTRACT

This paper presents four different hybrid genetic algorithms for network design problem in closed loop supply chain. They are compared using a complete factorial experiment with two factors, viz. problem size and algorithm. Based on the significance of the factor "algorithm", the best algorithm is identified using Duncan's multiple range test. Then it is compared with a mathematical model in terms of total cost. It is found that the best hybrid genetic algorithm identified gives results on par with the mathematical model in statistical terms. So, the best algorithm out of four algorithm proposed in this paper is proved to be superior to all other algorithms for all sizes of problems and its performance is equal to that of the mathematical model for small size and medium size problems.

Keywords: Closed Loop Supply Chain, Genetic Algorithms, HGA, Meta-Heuristics, MINLP, Model, Network Design, Optimization

1. INTRODUCTION

Supply chain management system has been being practiced as one of the main applications and research areas in the Industrial Engineering and Management discipline for the past two decades. A supply chain system is a complex network of business distribution channel which includes suppliers, manufacturers, distributors or wholesalers, warehouses, retailers, service providers and customers involved in the upstream and downstream flows of goods or services, information and finances. The effective and efficient management of the flows among these channel partners ensures the successive implementation of the entire supply chain. This can be achieved by adopting a coherent supply chain strategy with an appropriate network design, planning and implementation of the network for the execution of the flows in the supply chain system. The integration of entire supply chain flows into a closed loop network is the need of the hour now to ensure a business to be economically and environmentally sustainable with the changing trends in business and social environments, growing environmental consciousness in the society and government legislations to protect the environment as well as the business.

The integrated supply chain network design problems involve the decisions on the configuration of the network like, locating the right facilities, allocating the right capacity to the facilities, selection of the right mode of transportation and the route or path and optimizing the entire shipment of goods among the facilities or nodes in the entire distribution channel. These network design problems are often solved by using large optimization models. As these CLSC network design problems are combinatorial and complex in nature, handling of these problems with conventional optimization techniques like exact algorithms is quite difficult. Meta-heuristics algorithms like genetic algorithms and their hybrids are more suitable for these problems. In this context, this research work deals with a multi-echelon closed loop supply chain network design with forward and reverse logistics chains. In this research, an attempt has been made to develop a set of hybrid genetic algorithms and then select the best using through a complete factorial experiment.

The structure of this research paper is organized as follows. The next section gives the overview of the GAs and the third section deals with the literature review of the research works carried out in the past and recent periods and summarizes with the research gaps. The fourth section presents the statement of the problem identified and the fifth section presents the appropriate research methods adopted to solve the problem. The sixth section discusses the development and implementation of the mathematical model and the seventh section discusses the design, development and implementation of the hybrid genetic algorithms and its source codes. The eighth section discusses the comparison of the variants of hybrid genetic algorithms and determines the best HGA among them. The ninth section compares the best HGA with the mathematical model in terms of their performance measure for the optimization of the network. The last section concludes this research work with the summary of the main research outcome, scope and future research directions.

2. GENETIC ALGORITHMS—AN OVERVIEW

Genetic Algorithm (GA) is inspired by the theory of natural evolution and its principles. It is one of the directed random search techniques, employed to find a near-optimal solution for many larger problems in complex multi-dimensional search spaces. These algorithms encode a potential solution for complex problems on a simple chromosome like data structure and using techniques of natural evolution like inheritance, selection, mutation and crossover. After coding the solution in an appropriate way, GA iteratively works and evolves to get the global optimum without getting restricted at a local optimum. The individuals or chromosomes in a population are manipulated by the genetic operators to improve their fitness values while searching for global optimum solutions.

2.1. Stages of Genetic Algorithm

The steps involved in the formulation of Genetic Algorithms are mentioned as follows.

Chromosome Representation or Encoding: The individuals or the chromosomes are encoded or represented with a coding system to enable the computational processing of GA.

Initiation: Choose the initial population of the individuals.

Evaluation: Evaluate the fitness of each individual in population.

Breeding: Breeding is the process of reproduction with the following sub-steps which gets repeated until the termination condition gets satisfied.

Selection: Process of selecting the individuals with greater fitness for reproduction.

Crossover: Perform the crossover of selected individuals with a crossover probability for breeding new individuals which are offsprings.

Mutation: Perform the mutation process with a mutation probability on new individuals or offsprings for getting better offsprings.

Reevaluation: Evaluation the fitness of the each offspring created to find out the best or optimum fitness.

Termination: Terminate the process.

2.2. Classification of Genetic Algorithms

Genetic algorithms are theoretically and empirically proven to provide robust search and global solutions in complex scenarios for all sizes of problems. The Genetic Algorithms are capable of giving and efficient search in the problem domain and are finding wider range of applications in business, science, engineering and technology. These algorithms are more powerful in their search for improvement which has enabled the researchers to form different approaches of genetic algorithm. The genetic algorithms are classified as simple GA, parallel GA, hybrid GA, distributed GA, messy GA and adaptive GA etc., as given in Figure 1.

3. LITERATURE REVIEW

The literature is surveyed for the past 25 years and the varied research papers like concept papers, review papers, case studies, model papers and algorithmic research papers related to the principles, concepts of genetic algorithms, their application in the various closed loop and reverse supply chain network design problems are reviewed and presented in this section.

Goldberg and Deb [1] presented a comparative analysis of selection schemes used in Genetic Algorithms. They considered the commonly used selection schemes in GA viz. proportionate reproduction, ranking selection, tournament selection and Genitor (steady state) selection and compared them on the basis of solutions to deterministic difference or differential equations and verified it through computer simulations. Srinivas and Deb [2] presented their research on multi-objective optimization using non-dominated sorting in genetic algorithms. They investigated Goldberg's notion of non-dominated sorting in GAs along with a niche and speciation method to find multiple pareto-optimal points simultaneously. Golub [3] presented an implementation of binary and floating point chromosome representation in genetic algorithm. In both representations, the algorithm is based on steady-state reproduction, roulette-wheel bad individuals' selection and has the same parameters. The researcher concluded that the GA with floating point chromosome representation is simpler for implementation and it is faster than GA with binary representation. Deb [4] presented an introduction to Genetic Algorithm its concepts and scope of application. He presented the extension of simple GA and its population approach and ability to solve other search and optimization problems efficiently, including multimodal, multi-objective and scheduling problems, as well as fuzzy-GA and neuro-GA implementations. Lin et al. [5] proposed a generic scheme for adapting the crossover and mutation probabilities and presented an adaptive genetic algorithm for automatically adjusting suitable crossover and mutation rates to reduce the effort of searching for generation probability. Sim et al. [6] considered an extended supply chain network design model with reverse logistics and proposed an LP-based genetic algorithm that uses the LP-based solution and genetic operators. Yeh [7] presented a hybrid heuristic algorithm for the multistage supply chain network problem. The researcher employed a simple greedy method and a hybrid local search method combining the linear programming technique (LP), the pairwise exchange procedure (XP), the insert procedure (IP) and the remove procedure (RP) to solve the MSCN problem, in the proposed algorithm.

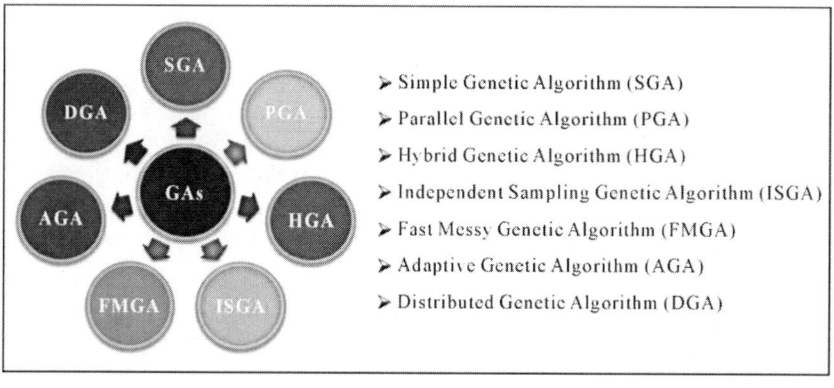

Figure 1. Classification of genetic algorithms.

Altiparmak et al. [8] presented a mixed integer non-linear programming model for multi-objective optimization which considers the three objectives: 1) minimization of total cost comprised of fixed costs of plants and distribution centers (DCs), inbound and outbound distribution costs, 2) maximization of customer services that can be rendered to customers in terms of acceptable delivery time(coverage), and 3) maximization of capacity utilization balance for DCs (i.e. equity on utilization ratios). They implemented two different weight approaches (i.e. GA_A1 and GA_A2) in the proposed GA at the first stage to evaluate the alternate solutions and compared the Pareto optimal solutions obtained with that of the multi-objective simulated annealing (MO_SA) at the second stage. It was found that GA_A1 outperformed the MO_SA in terms of the quantity and quality of the Pareto optimal solutions obtained. Min, et al. [9] dealt with the reverse logistic problem involving spatial and temporal consolidation of returned products in a closed-loop supply chain network. They proposed a mixed-integer, nonlinear programming model and a genetic algorithm to solve the problem. El-Mihoub et al. [10] reviewed the Hybrid Genetic Algorithms. The different forms of integration between genetic algorithms and other search and optimization techniques are presented. They discussed that hybridization has been utilized to construct competent genetic algorithms that belong to two of the three main approaches for building competent algorithms, i.e., perturbation, linkage adaptation, and probabilistic model-building. Zhou et al. [11] proposed a mixed integer non-linear programming model and a hybrid genetic algorithm to solve the supply chain distribution problem with uncertain demands and product returns, simultaneously. The proposed HGA was verified with a series of computational experiments and showed efficiency in obtaining a near-optimal solution for the distribution system problem. Schultmann et al. [12] presented modeling of reverse logistics tasks within closed loop supply chains based on an example considering the end of life vehicle (ELV) treatment practiced in the automobile industry of Germany. They proposed different design options for a closed loop supply chain concentrating on the handling of the reverse material flows to reintegrate them into their genuine supply chains. They modeled the reverse logistics aspects with vehicle routing planning and developed a problem-tailored algorithm.

Altiparmak et al. [13] proposed a steady state genetic algorithm (ssGA) with a new encoding structure and a new greedy heuristics for the single-source, multi-product, multi-stage supply chain network design problem. They investigated the effectiveness the ssGA by comparing the results with those obtained by CPLEX, Lagrangean heuristics, hybrid GA and simulated annealing on two sets of test problems with different sizes. Lin, et al. [14] presented flexible multistage logistics network (fMLN) design with location allocation problems. They proposed an effective hybrid genetic algorithm to solve the problem, by adding some guiding information into the chromosome of the existing pGA in the literature and also employed local optimization based decoding method and proposed a new combinatorial crossover to improve the performance of rpGA. Staikos and Rahimifard [15] presented a decision-making model for waste management in the footwear industry through reverse logistics model which investigated into the steps required to consider the end-of-life implication of shoes and promote post-consumer recycling practices in the industry. In the product recovery process, they included the recycle, repair/reuse, land-filling and incineration operations to minimize the cost and EOL wastage for an environmentally sustainable network model. Ko and Evans [16] developed an optimization model and associated algorithm to design an integrated logistics network for 3PL providers with the simultaneous flows of forward and reverse logistics network and presented a mixed integer nonlinear programming model which is a multi-period, two-echelon, multi-commodity, capacited network design problem. They proposed a GA-based heuristic with genetic operations and transshipment algorithm to solve the problem.

Min and Ko [17] presented a proposed a mixed-integer programming model and a genetic algorithm for solving a reverse logistics design problem which includes the location and allocation of repair facilities for 3PLs. The researchers suggested the comparison of proposed GA with other heuristics like Lagrangian relaxation and Tabu search methods for future investigations. Belgasmi et al. [18] considered a multi-location transshipment model with limited storage capacity with an objective to minimize the aggregate cost function where decision variables are the constrained up-to-date quantities. The researchers proposed a real coded genetic algorithm (RCGA) with a new crossover operator to approximate the optimal solution. The GA proved its ability to solve instances of the problem with high accuracy. Farahania and Elahipanaha [19] developed a new model in a multi-objective mixed integer linear programming for a three-echelon supply chain network to minimize the total costs and service level for just-in-time distribution purposes for the real world situations. They developed a genetic algorithm (GA) to find the pareto optimal solutions for the large sized instances of this problem. Lee and Don [20] developed a mathematical model for the integration of forward and reverse distribution network design and the locations of facilities jointly used by forward and reverse logistics operations. They developed a two-stage heuristic algorithm as the first attempt in solving the integrated forward and reverse logistics network design problem using meta-heuristics. A tabu search is also applied to obtain the improved solution of shipping the returned products.

Yun et al. [21] designed a multi-stage supply chain model and proposed a hybrid genetic algorithm (hGA) with adaptive local search scheme which determines the usage of local search technique in GA loop. They suggested two multi-stage based supply chain models with complicated routes and compared the proposed hGA with other conventional algorithms like GA, enumeration method, using numerical examples for their performances. The proposed hGA outperformed the competing algorithms. Sourirajan et al. [22] presented a two stage supply chain with a production facility that replenishes a single product retailer with an objective to locate distribution centers in the network such that the sum of facility location, pipeline inventory, and safety stock costs is minimized. They used genetic algorithm to solve the model and compared the performance to that of a Lagrangian heuristic in earlier work. They applied a new chromosome representation that combines binary vectors with random keys provides solutions of similar quality to those from the Lagrangian heuristic. Lee et al. [23] proposed a model for optimization of reverse logistics network using hybrid genetic algorithm. They proposed a genetic algorithm with priority-based encoding method consisting of 1st and 2nd stages combined a new crossover operator called as weight mapping crossover (WMX). They applied a heuristic in the 3rd stage to transportation of parts from processing center to manufacturer.

Gen et al. [24] dealt with bi-criteria network design (bND) problem using multi objective hybrid genetic algorithm (mo-hGA) to minimize the total cost and maximize the total flow. They proposed a new chromosome representation based on priority based encoding method and a new crossover operator called as weight mapping crossover (WMX), and adopted insertion mutation operator and hybrid approach by fuzzy logic control (FLC) and local search (LS). Lin et al. [25] formulated an integrated multi-stage logistics network model by considering the direct shipment and direct delivery of logistics and inventory. A mixed integer programming model is formulated to to minimize the sum of transportation cost, inventor holding cost, fixed ordering cost and open cost of facilities. An effective hybrid evolutionary algorithm is proposed to solve the location-allocation problem combined with lot sizing problem which is a more complex and NP hard in nature. Salema et al. [26] presented a strategic and tactical model for closed loop supply chain. They integrated the strategic network design decisions with the tactical decisions like production, storage and distribution planning and achieved the integration by considering the micro and macro time scales. They formulated a mixed integer linear programming and solved the model using standard branch & bound techniques.

Costa et al. [27] proposed a new efficient encoding-decoding procedure for chromosome representation in genetic algorithm to minimize the total logistic cost resulting from the transportation of goods and the location and opening of the facilities in a single product three-stage supply chain network. The procedure is based on a parsimonious permutation decoding of the string representing the network and a three steps decoding procedure allowing the occurrence of unfeasible transportation trees to be reduced. Pishvaee et al. [28] proposed a model for integrated logistic network design to address the issues due to increasing network costs and network responsiveness in supply chain and reverse logistics. They presented a bi-objective mixed integer non-linear programming (MINLP) model for the integrated forward and reverse logistics network design. The researchers designed a multi-objective memetic algorithm with dynamic local search mechanism (MOMA) to solve the model with non-dominated set of solutions. On comparison with LINGO, MOMA obtained a reasonable quality of solutions on the multiple capacity test problems.

Zarei et al. [29] designed a conceptual model of a logistics network including forward and reverse chains for the production of new vehicles and recovery of used ones. They developed a mathematical model to minimize the cost of setting up the network and also the relevant transportation costs. They designed a genetic algorithm to deal with the complexity of the problem. Kannan et al. [30] presented the case of a battery recycling in which a closed loop mixed integer linear programming model was developed to determine the raw material level, production level, distribution and inventory level, disposal level, and recycling level at different facilities with the objective of minimizing the total supply chain costs. The researchers proposed heuristics based genetic algorithm (GA) for solving the problems and computational results obtained through GA are compared with the solutions obtained by GAMS optimization software. They found that for smaller size problems the GAMS software provides better results but with worst computational time and also found that some larger-size real-world problems which cannot be solved by GAMS or other commercial software are only solved by the proposed heuristics based GA. The result comparison between GA and GAMS shows that the heuristics based GA as a solution methodology is reliable for the larger problem sizes. El-Sayed et al. [31] developed a multi-period, multi- echelon forward-reverse logistics network design under risk. The network structure includes three echelons in the forward direction(suppliers, facilities and distribution centers) and two echelons in the reverse direction (disassembly and redistribution centers), first customer zones in which the demands are stochastic and second customer zones in which the demand is assumed to be deterministic. They formulated the problem in a stochastic mixed integer linear programming (SMILP) decision making form as a multi-stage stochastic program to maximize the total expected profit.

Kaya et al. [32] presented a novel crossover operator for genetic algorithms called ring crossover. In order to evaluate the efficiency and feasibility of the proposed operator, a comparison between the results of this study and results of different crossover operators used in GAs is made through a number of test functions with various levels of difficulty. The most important advantage of the proposed method is that more variety is presented in possible number of children according to SPC and TPC operators. Khajavi et al. [33] presented an integrated forward/reverse logistics network optimization model for multi-stage capacited supply chain network. The model is proposed by formulating the generalized logistics network problem into a bi-objective mixed-integer programming model to minimize the total costs and maximize the responsiveness of the CLSC network simultaneously. The proposed model was able to integrate the forward and reverse logistics network design decisions to avoid the sub-optimality resulted from separated and sequential designs. Nandita [34] proposed a reverse logistics model for the apparel product acquisition via two channels viz., used garment collectors and used garment importers in the Indian apparel market. The author also presented the recovery process of the used garments as per their value potential and the level of reconditioning required. Further, reconditioning, reconstruction and recycling processes in the reverse logistics chain for the redistribution and sales in the apparel aftermarket in India have been examined.

Hosseinzadeh and Roghanian [35] presented an optimization model for reverse logistics network under stochastic environment using genetic algorithm. The researchers developed a probabilistic mixed integer linear programming model for multi-product, multistage reverse logistics network design problem which considers minimizing of total shipping cost and fixed opening costs of the disassembly centers and the processing centers in the reverse network. A priority based genetic algorithm was applied to solve the problem using a numerical example. Kumar and Jyotishree [36] studied different encoding techniques and their genetic operations and proposed a new encoding scheme to overcome the limitations of existing encoding techniques. The proposed encoding scheme is independent of order or value, represents genes in the chromosome as its fitness contribution value and the order of genes is immaterial. The encoding scheme supports one point crossover as in binary encoding and PMX crossover as in permutation encoding. The proposed encoding scheme is also apt for inversion operation which is restricted to some encoding schemes. Bozorgirad et al. [37] proposed Route Based Genetic Algorithm (RBGA) to find the minimum cost of flexible multi stage logistic network.

They considered a multi- source multi product flexible multistage logistic network and presented the comparison of numerical results of RBGA and standard GA. They applied the penalty method in GA and new representation of GA to satisfy all existing constraints.

Mehdizadeha and Afrabandpeia [38] presented a multi-stage and multi product logistic network design problem for which they formulated a mixed integer nonlinear programming model (MINLP) to minimize the transportation and holding costs. Then they developed a hybrid priority based genetic algorithm (pb-GA) and simulated annealing algorithm (SA) in two phases. The researchers determined the amount of products to be carried between the sources and depots in the first phase and determined the vehicles for transporting products in the second phase. The researchers used Response Surface Methodology (RSM) to adjust the significant parameters of the algorithm. Iris and Serdarasan [39] presented the applications of genetic algorithms to solve the supply chain network design problem. The researchers introduced different GA approaches that are used at different steps of the algorithm, such as encoding/decoding techniques, initialization strategies, fitness function determination, selection techniques, crossover and mutation strategies. Zaki et al. [40] presented an efficient algorithm for solving multi-objective transportation, assignment and transshipment problems. The researchers proposed an integrated approach of both genetic algorithm (GA) and local search (LS) scheme. The algorithm is an iterative multi-objective genetic algorithm with an external population of pareto optimal solutions that best confirm a Pareto front. Then the algorithm implements GA to provide the initial set (close to the pareto set as possible followed by local search method to improve the quality of the solutions. Ozkir and Basligil [41] proposed modeling of product recovery processes in closed loop supply chain network design, where a mixed integer linear programming model was applied to obtain CLSC network design in which the recovery process occurs as material recovery, component recovery and product recovery.

Rafsanjani and Eskandari [42] proposed a genetic algorithm (GA) approach with segment based operators combined with a local search technique (SHGA) to solve the multi stage supply chain network design problem. The researchers evaluated the performance of the proposed algorithm (SHGA) by comparing the results with that of several other algorithms like simulated annealing (SA), a simple genetic algorithm without segment based operators (GA), hybrid genetic algorithm with local search and without segment based operators (HGA) and also with a proposed segment based GA without the local search (SGA) on different MSCN design problems. Lee et al. [43] considered single product and two stage reverse logistics problem with two objectives of minimizing the total transportation cost (includes fixed opening cost, transportation cost and inventory cost) and minimizing the total tardiness in all time periods. The researchers proposed a multi objective hybrid genetic algorithm (mo-hGA) combined with Fuzzy Logic Controller (FLC) for efficiently dealing with multi-objective logistics (mo-RLN) problem. They applied a priority based chromosome representation and adaptive weight approach in the hybrid GA for solving the problem. Soleimani et al. [44] proposed a multi-echelon multi-period and multi product CLSC model using genetic algorithm. They solved the proposed model using CPLEX optimization software and by a developed genetic algorithm. They compared the results of proposed pre-tuned genetic algorithm with a global optimum of CPLEX solver. Then they generated a sufficient large number of large-size instances and solved them by the proposed GA. The results proved the constancy and acceptable performances of the proposed GA and its applicability in real-size situations. Lee and Lee [45] developed a closed loop supply chain model for saving the cost of integrating the forward and reverse supply chains with by considering the uncertainties in demand and supply in forward chain, and uncertainties in the time frame of recovery and the amount of recovered materials in reverse chain. The optimal costs are determined by applying an optimization algorithm using the priority based genetic algorithm and the modified genetic algorithm. This modified hybrid genetic algorithm (mhGA) provides an improved search-ability of best solution and the convergence speed. Teodoro et al. [46] proposed a new hybrid operator for genetic algorithm used to optimize the purchase of products in stores geographically separated in order to obtain solutions that combine the purchase of products with the lowest possible total cost, considering the route between each store. They described the development of a strategy based on GA, emphasizing the proposition of a new hybrid genetic crossover operator, to provide a set of optimized solutions for a SCM specific problem. The proposed operator has proven to be effective in situations where the aim is to ensure solutions with high genetic diversity, but with the same best fitness value. For problems where the goal is to achieve the best solution, the two-point and/or uniform crossover can be used. Ramezani et al. [47] presented a new multi-objective stochastic model for a forward/reverse logistic network design with responsiveness and quality level. They included three echelons in forward flow viz. suppliers, plants and distribution centers and two echelons in the backward flow viz. collection centers and disposal centers. They evaluated the systematic supply chain configuration maximizing the profit, customer responsiveness and quality to achieve the objectives of the network. Rosa et al. [48] presented a robust sustainable bi-directional logistics network design under uncertainty which handles a network of multiple supply stages, including production allocations, uncertain data development, facility locations and flexible capacity adjustments. They first introduced a detailed deterministic model assessing the impact of incorporating reverse logistics into a forward-oriented supply chain then extended it to a robust capacited facility location model, which minimizes the expectations of relative regrets for a set of scenarios over a multi-period planning horizon, by considering uncertainty in supplying and collecting goods.

Mahmoudi et al. [49] proposed mathematical modeling for minimizing costs in a multi-layer, multi-product reverse supply chain. They presented an integer linear programming model for multi-layer, multi-product reverse supply chain that minimizes the products and parts transportation costs among centers and also sites launch, operation parts, maintenance and remanufacturing costs at the same time. The model was solved and validated using LINGO 9 software. Hafeti and Jolai [50] proposed a robust and reliable network design for forward-reverse logistics design which simultaneously takes care of uncertain parameters and facility disruptions in the network. They proposed a mixed integer linear programming model with augmented p-robust constraints considered to control the reliability of the network during disruptions thereby reducing the nominal cost and disruptions risks. They compared the performance of the augmented p-robust criterion with that of the other conventional robust criteria. Cardoso et al. [51] developed a mixed integer linear programming for the design and planning of supply chains with reverse flows by simultaneously considering the production, distribution and reverse logistics activities for which the product demand is uncertain. The model defines the maximization of the expected net present value and provides details on sizing and location of plants, warehouses and retailers, definition of processes to install, establishment of forward and reverse flows and inventory levels to achieve.

Devika et al. [52] presented the design of a sustainable closed loop supply chain network to cover the gap in the quantitative modeling by considering the social impacts, environmental impacts and economic impacts in the network design problem. They developed three new hybrid meta-heuristic methods based on adapted imperialist competitive algorithms and variable neighbourhood research, to solve this NP hard problem. They compared the algorithms with each other and also with other strong algorithms to assess the efficiency and effectiveness of the proposed algorithms. Asghari and Nezhadali [53] considered a design of reverse logistics network development for recycling waste products to reduce harmful emissions due to storage, transportation and various processes. They proposed a multi objective model using mixed integer linear programming and solved the problem by applying a hybrid method based on non-dominated sorting genetic algorithm (NSGA II). Aggarwal et al. [54] discussed the general principles of Genetic algorithm, its structure and mechanism in their paper.

They presented different encoding schemes like binary encoding, permutation encoding, value encoding and octal encoding schemes used in genetic algorithm. Demirel et al. [55] proposed a mixed integer programming model for a closed loop supply chain (CLSC) network with multi-periods and multi-parts under two main policies as secondary market pricing and incremental incentive policies. The researchers developed a fuzzy multi-objective extension in addition to the base case (crisp) formulation to solve CLSC network problem with fuzzy objectives to represent vagueness in real-world problems. A genetic algorithm (GA) approach was applied to solve the real size crisp and fuzzy CLSC problems. The researchers investigated and illustrated the effectiveness of the proposed meta-heuristic approach by comparing the results with GAM-CPLEX on a set of crisp/fuzzy problems with different sizes. Dzupire and Gyeke [56] developed a multi-stage supply chain network optimization using genetic algorithms. The developed model minimizes system wide costs of the supply chain and delays on delivery of products to distribution centers for a three echelon supply chain. The researchers used NSGA-II through the Global Optimization Toolbox in MATLAB. The algorithm clearly provides the Pareto fronts which are efficient solutions for improving the supply chain design problems. The researchers proved that the proposed multi objective evolutionary algorithms are successful in dealing with multi-objective optimization problems and have potential to solve combinatorial problems. Rad et al. [57] formulated a bi-criteria multi source single product fMLN model that considers the minimization of the total transportation cost and total product delivery time simultaneously. A meta-heuristic technique i.e. genetic algorithm with an enhancement to solve the complex fMLN is proposed. A set of pareto optimal solution is obtained using new chromosome representation capable of improving the constraints of the problem by a considerable ratio and with the defined crossover and mutation techniques. The results showed that the solutions of the proposed GA are better than that of the standard GA. Aravendan and Panneerselvam [58] considered a multi-echelon closed loop network design problem with an objective to minimize the total cost of the network and increasing the quality of service. They developed a mixed integer non-linear programming model to solve the CLSCND problem. The researchers simulated the experiments and solve the model using the optimization software LINGO14.

Sarrafhaa et al. [59] presented a bi-objective integrated procurement, production and distribution problem of a multi-echelon supply chain network design in order to minimize the total cost of the supply chain and the average tardiness of the products delivered to DCs by factories. The authors developed a multi periodic structure in the network design involving suppliers, factories, distribution centers (DCs) and retailers. A novel Pareto based algorithm MOBBO was developed to find pareto optimal solutions. They validated the results using existing NSGA-II and MOSA, where the parameters of all algorithms were tuned using the Taguchi method. The results showed that the proposed MOBBO is a compatible alternative to the competing algorithms. Pasandideh et al. [60] presented a bi-objective optimization of a multi-product multi-period three echelon supply chain under uncertain environments. The researchers first formulated the problem into the framework of a single objective stochastic mixed integer linear programmimg model (MILP) then reformulated the problem into a bi-objective deterministic mixed integer nonlinear programming model (MINLP). They used a non-dominated sorting genetic algorithm (NSGA-II) to solve the problem and applied another GA based algorithm called non-dominated ranking genetic algorithm (NRGA) to validate the results obtained. A modified priority based encoding is proposed in both algorithms. The researchers showed the applicability of the proposed method by providing numerical illustrations and applied t test along with the simple additive weighting (SAW) method to select the best method. The results showed that NSGA-II acted better than NRGA for all problem instances.

3.1. Review Summary

The literature is reviewed for the design of reverse supply chain systems as well as the closed loop supply chain systems i.e. the reverse supply chain networks in conjunction with forward supply chain networks under the following nine categories, viz. models, branch and bound algorithms, heuristics, genetic algorithms, simulated annealing algorithms, Petri net algorithms, simulation approaches and general approaches. The review outcome shows that the Meta heuristics particularly the Genetic Algorithm and its hybrids were applied majorly by the researchers to deal with the supply chain network design problems. The models developed were majorly based on MILP and a few were also on MINLP, depends on the complexity of the problems. The research gaps with respect to modeling and algorithmic approaches are identified and furnished as follows.

3.2. Research Gaps with Respect to Model

Most of the researches carried out in the prior periods deal with the CLSCND model very specific to a supply chain problem of a particular industry or a product category or a type of distribution network. There is a less or lack of research on the generic integrated closed loop network design model which can be applicable to a majority of the Industries or to the vast range of product categories or all kinds of distribution network. So, there is a need for the generic integrated model proposed in this research. This integrated model can be applicable to majority of the industries or product category or any kind of distribution network just by making minor addition or deletion in the stages or levels or echelons involved in the integrated model. Thus, this model can be applicable to various industrial sectors like Automobiles, Leather products, Textile products, Footwear, Consumer Electronics, Home appliances and Industrial equipment or machines, etc.

3.3. Research Gaps with Respect to Algorithms

The past researches in the supply chain network design deal with the development of heuristics / meta heuristics mostly for forward chain network designs. A fair amount of research has been carried out to develop heuristics for reverse supply chain network designs. A very less research has been done to develop heuristics /Meta heuristics for the closed loop supply chains which are gaining importance in recent times. Also most of genetic algorithms developed in the past were designed with a specific set of GA parameters. A very less research has been done to study the performance of the GA with respect to changing GA parameters like selection methods, cross over methods, parent replacement methods etc., for the same problem. This gap is addressed in this research by developing new variants of hybrid genetic algorithm with different combination of GA parameters like using Elitism & Rank Selection methods, Chromosome Cross over techniques and Chromosome Replacement techniques.

4. STATEMENT OF THE PROBLEM

The closed loop supply chain network model considered in this research work is a multi-echelon and multi stage network which includes manufacturers, wholesalers, retailers and first customers in the forward chain and service/repair centers, collector/dismantler/re-furbisher (CDR), remanufacturers, recyclers, disposal centers/land- fillers, resellers and second customers in the reverse chain. The network deals with the two types of product returns i.e. product returns due to repair and product returns due to end of use or life, in its reverse chain. The repair products are sent directly to the repair/service centre by the first customer for getting them repaired and the end of life products are returned either directly or through the retailer to the collector/dismantler/re-furbisher by the first customer for re-processing. The returned products thus collected in the CDR locations are sorted; parts are dismantled and segregated into recoverable and non-recoverable or waste items. The recoverable items are again segregated into re-furbishable, remanufacturable, recyclable items and shipped to the respective facilities or locations for recovery process. The non-recoverable or the waste items are shipped to the disposal centers/ Land-fillers and disposed through land filling or incineration. The recovered products through the process of refurbishing and remanufacturing are shipped to the second customers via resellers. The items recovered through recycling process are shipped to the raw material suppliers market by the recycler. Based on these discussions, it is clear that the returned products due to EOL are shipped from the first customers to the recovery facilities through a push mechanism and recovered products are shipped from the recovery facilities to the second customers through a pull logistics mechanism. In this context, the objective of this research work is to minimize the total cost which is the sum of the costs of both forward and reverse supply chains there by increasing the total profit of the closed loop network. After the text edit has been completed, the paper is ready for the template. Duplicate the template file by using the Save As command, and use the naming convention prescribed by your journal for the name of your paper. In this newly created file, highlight all of the contents and import your prepared text file. You are now ready to style your paper.

5. RESEARCH METHODS

The research methods adopted for solving the research problem stated in the previous section are Modeling and Algorithmic research methods which are briefed as follows.

Modeling Research: A model is an abstraction of reality applied to the real life problems of business situations. Among the three types of modeling research methods, viz. symbolic model, mathematical model and simulation model, mathematical modeling research method is applied to design and develop the appropriate model for the specific CLSCND problem. The model is based on the Multi-Integer Non-Linear Progrmme (MINLP) as the problem is combinatorial and NP Hard in nature.

Algorithmic Research: An algorithm is a well-defined sequence of steps to solve a problem of interest in industry, business, etc. Exact algorithms, heuristics/meta-heuristics are applied to solve these problems based on whether the problem is polynomial or combinatorial in nature. In this research work, since the problem is combinatorial, complex and NP hard in nature, a Meta-heuristic i.e. Genetic Algorithm is applied to design and develop the appropriate algorithm for the specific CLSCND model. A Genetic Algorithm with hybrid architecture is planned due to its:

- Global search nature leading to more accurate global optimal solutions rather than local optimal solutions by exact algorithms/heuristics;
- Random pick up of chromosomes from the wider population and longer generations;
- Faster in solving the problems and efficient in dealing the combinatorial NP hard problems.

6. DEVELOPMENT OF MATHEMATICAL MODEL (ARAVENDAN AND PANNEERSLVAM [58])

Aravendan and panneerslvam [58] proposed a model for the closed loop network design problem which integrates both forward and reverse logistics in the supply chain. The closed loop network presented in their research paper is a single product, single period and multi-echelon supply chain which channelizes manufacturers, wholesalers, retailers and first customers in the forward chain and channelizes repair/service centers, collectors/ dismantlers/re-furbishers, remanufacturers, recyclers, land fillers, resellers and second customers in the reverse chain.

In the closed loop network design model, in the forward supply chain, the manufacturers are responsible for manufacturing new or virgin products and supplying them to the wholesalers for distribution. The wholesalers are responsible for the distribution of new products to the retailers in their region. The retailers are responsible for selling the new products to the first customers as per their demands and also responsible for facilitating the after sales service. The customers' nodes represent one or more customers or a group of customers. The first customers are responsible to return the products supplied to them as per the demand, either during the usage or after the usage of the products as either repair product returns or end of life product returns (EOL) respectively. In the reverse supply chain, the first customers return the repair products to the service/repair centre for getting them repaired and to reuse. The repair/service centers are responsible for providing the quality service to the customers and to ensure the prompt delivery of the repaired products to the first customers for reuse. The end of life products are returned to the collector/dismantler/re-furbisher (CDR) either directly or via retailers by the first customers. These

CDR locations are responsible for collecting, dismantling and sorting the returned products and dismantled parts for refurbishing, remanufacturing, recycling and disposing via landfill or incineration. They are also responsible for supplying the remanufacturable to the manufacturers, recyclable to the recyclers, disposables to the land-fillers/incinerators. The CDR locations recondition the refurbishable products and distributing them directly to the resellers. The resellers also receive the remanufactured products from the remanufacturers and they sell both remanufactured and refurbished products to the second customers as per their demands. The recyclers are responsible for recycling the recyclable items received from CDR locations. The disposal centers/land-fillers are responsible for the safe disposal of the unusable wastes received from CDR locations either by landfilling or incinerations.

The assumptions and limitations of the proposed model are considered as follows: 1) The model is for a single product and single period network design, 2) The locations of the first customers and second customers are known and are with certain demands, 3) The quantities of products returned are certain and all the products supplied are returned as EOL products and repair products, 4) 60% of the products supplied are returned as EOL products and 40% are returned as repair product, 5) 50% of the EOL products are returned via retailers and remaining 50% of the EOL products are returned directly, to the Collectors/Dismantlers/Re-furbishers (CDR), 6) Out of the total returned EOL products, 30% are re-furbishable items, 45% are re-manufacturable items, 20% are recyclable items and 5% are non-recoverable and disposed by land-filler, 7) The quality of the remanufactured, refurbished and repaired products is different from that of the new product, 8) The potential locations of manufacturers, wholesalers, retailers, collectors/dismantlers/re-furbishers, repair/service centers, recyclers, land fillers and resellers are assumed, 9) The capacity of each location is known, 10) The costs parameters considered (viz., opening costs, operating costs, un-utilized capacity costs and transportation costs) are known for all the facilities and node, 11) The measure of quantity of products transported per trip is defined in the form of number of units per trip and 12) There is no shipment happening between the nodes in the same stage.

6.1. Costs Associated with the Closed Loop Supply Chain Network Design Model

The various costs incurred at different nodes in the multi-echelons of the closed loop supply chain network are opening costs, operation costs and transportation costs for manufacturers, wholesalers and retailers in the forward chain, and the same costs are applied for repair/service centers, collectors/dismantlers/re-furbishers, land fillers, recyclers, remanufacturers and resellers in the reverse chain. An un-utilized capacity cost is also considered and added only to the hybrid centers manufacturers/remanufacturers. There is no cost considered for first customers and second customers as the products are picked up by them at the retail points in this particular model.

6.2. The Mathematical Model

The proposed model has the objective of minimizing the total cost i.e. the cost of the shipment flows in both forward and reverse supply chains as given in Figure 2 for which the objective function is defined and described as follows:

$$\begin{aligned}
\text{MIN } Z_{\text{Total Cost}} &= (\text{Total Cost}) \\
&= (\text{Opening Cost} + \text{Operation Cost} + \text{Un} - \text{Utilized Capacity cost} + \text{Transportation Cost})
\end{aligned}$$

$$\begin{aligned}
\text{MIN } Z_{\text{Total Cost}} &= \sum_{i=1}^{I} a_i \times Y1_i + \sum_{j=1}^{J} b_j \times Y2_j + \sum_{k=1}^{K} c_k \times Y3_k + \sum_{m=1}^{M} e_m \times Y5_m + \sum_{n=1}^{N} f_n \times Y6_n + \sum_{i=1}^{I} ar_i \times Y1_i \\
&+ \sum_{o=1}^{O} g_o \times Y7_o + \sum_{p=1}^{P} h_p \times Y8_p + \sum_{r=1}^{R} u_r \times Y9_r + \sum_{i=1}^{I} (Q0_i \times Y1_i) \cdot OC1_i + \sum_{i=1}^{I} \sum_{n=1}^{N} (Q12_{ni} \times Y1_i) \cdot OCR1_i \\
&+ \sum_{i=1}^{I} \sum_{j=1}^{J} (Q1_{ij} \times Y2_j) \cdot OC2_j + \sum_{k=1}^{K} \left\{ \sum_{j=1}^{J} (Q2_{ik} \times Y2_i \cdot Y3_k) + \sum_{l=1}^{L} (Q3_{kl} \times Y3_k) + \sum_{l=1}^{L} (Q7_{lk} \times Y3_k) \right\} \cdot OC3_k \\
&+ \sum_{l=1}^{L} \sum_{m=1}^{M} (Q5_{ml} \times Y5_m) \cdot OC5_m + \sum_{n=1}^{N} \left\{ \sum_{l=1}^{L} (Q6_{ln} \times Y6_n) + \sum_{k=1}^{K} (Q8_{kn} \times Y3_k \cdot Y6_n) \right\} \cdot OC6_n \\
&+ \sum_{o=1}^{O} \sum_{n=1}^{N} (Q9_{no} \times Y6_n \cdot Y7_o) \cdot OC7_o + \sum_{p=1}^{P} \sum_{n=1}^{N} (Q10_{np} \times Y6_n \cdot Y8_p) \cdot OC8_p \\
&+ \sum_{i=1}^{I} \sum_{r=1}^{R} (Q13_{ir} \times Y1_i \cdot Y9_r) + \sum_{n=1}^{N} \left\{ \sum_{r=1}^{R} (Q11_{nr} \times Y6_n \cdot Y9_r) + \sum_{r=1}^{R} \sum_{r=1}^{R} (Q14_{rs} \times Y9_r) \right\} \cdot OC9_r \\
&+ \sum_{i=1}^{I} \sum_{j=1}^{J} \left[CPM_i - (Q1_{ij} \times Y1_i \cdot Y2_j) \right] \cdot UC1_i + \sum_{i=1}^{I} \sum_{r=1}^{R} \left[CPRM_i - (Q13_{ir} \times Y1_i \cdot Y9_r) \right] \cdot UCR1_i \\
&+ \sum_{i=1}^{I} \sum_{j=1}^{J} (Q1_{ij} \times D1_{ij} \times TC) Y1_i \cdot Y2_j + \sum_{j=1}^{J} \sum_{k=1}^{K} (Q2_{jk} \times D2_{jk} \times TC) Y2_j \cdot Y3_k \\
&+ \sum_{k=1}^{K} \sum_{l=1}^{L} (Q3_{kl} \times D3_{kl} \times TC) Y3_k + 2 \sum_{l=1}^{L} \sum_{m=1}^{M} (Q4_{lm} \times D4_{lm} \times TC) Y5_m + \sum_{l=1}^{L} \sum_{n=1}^{N} (Q6_{ln} \times D6_{ln} \times TC) Y6_n \\
&+ \sum_{l=1}^{L} \sum_{k=1}^{K} (Q7_{lk} \times D7_{lk} \times TC) Y3_k + \sum_{k=1}^{K} \sum_{n=1}^{N} (Q8_{kn} \times D8_{kn} \times TC) Y3_k \cdot Y6_n \\
&+ \sum_{n=1}^{N} \sum_{o=1}^{O} (Q9_{no} \times D9_{no} \times TC) Y6_n \cdot Y7_0 + \sum_{n=1}^{N} \sum_{p=1}^{P} (Q10_{np} \times D10_{np} \times TC) Y6_n \cdot Y8_p \\
&+ \sum_{n=1}^{N} \sum_{i=1}^{I} (Q12_{ni} \times D12_{ni} \times TC) Y6_n \cdot Y1_i + \sum_{i=1}^{I} \sum_{r=1}^{R} (Q13_{ir} \times D13_{ir} \times TC) Y1_i \cdot Y9_r \\
&+ \sum_{n=1}^{N} \sum_{r=1}^{R} (Q11_{nr} \times D11_{nr} \times TC) Y6_n \cdot Y9_r + \sum_{r=1}^{R} \sum_{s=1}^{S} (Q14_{rs} \times D14_{rs} \times TC) Y9_r
\end{aligned}$$

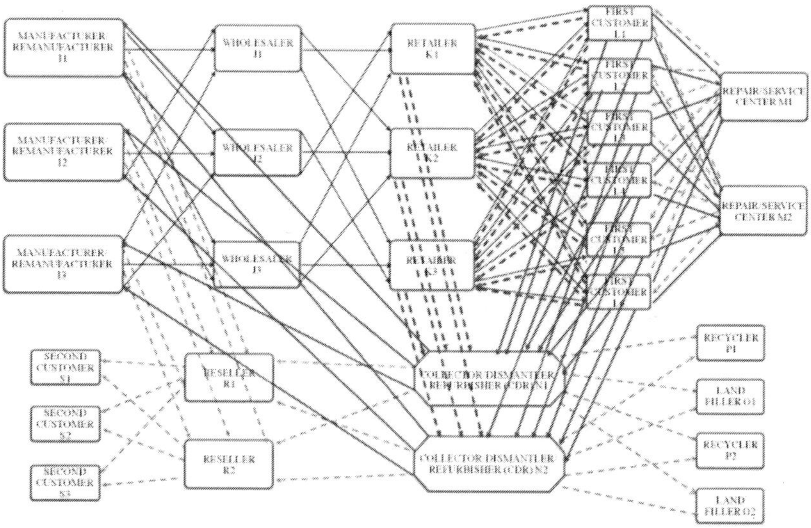

Figure 2. Shipment flows of CLSCND before optimization.

Subject to

Demand Constraints (DC):

$$\sum_{k=1}^{K} Q3_{kl} \times Y3_k \geq \sum_{l=1}^{L} D1_l, \text{ where } l = 1,2,3,\cdots,L, \text{ First Customer Demand Constraints} \tag{1}$$

$$\sum_{r=1}^{R} Q14_{rs} \times Y9_r \geq \sum_{l=1}^{L} D2_s, \text{ where } s = 1,2,3,\cdots,S, \text{ Second Customer Demand Constraints} \tag{2}$$

Capacity Constraints (CC):

$$\sum_{j=1}^{J} Q1_{ij} \times Y1_i \cdot Y2_j \leq CPM_i, \text{ where } i = 1,2,3,\cdots,I,\ j = 1,2,3,\cdots,J \tag{3}$$

$$\left[\sum_{j=1}^{J} Q1_{ij} \times Y1_i \cdot Y2_j + \sum_{r=1}^{R} Q13_{ir} \times Y1_i \cdot Y9_r - \sum_{n=1}^{N} Q12_{ni} \times Y6_n \cdot Y1_i\right] \leq CPM_i,$$
$$\text{where } i = 1,2,3,\cdots,I,\ j = 1,2,3,\cdots,J,\ n = 1,2,3,\cdots,N,\ r = 1,2,3,\cdots,R \quad (4)$$

$$\sum_{k=1}^{K} Q2_{jk} \times Y2_j \cdot Y3_k \leq CPW_j, \text{ where } j = 1,2,3,\cdots,J,\ k = 1,2,3,\cdots,K \quad (5)$$

$$\left[\sum_{l=1}^{L} Q3_{kl} \times Y3_k + \sum_{n=1}^{N} Q8_{kn} \times Y3_k \cdot Y6_n\right] \leq CPRT_k,$$
$$\text{where } k = 1,2,3,\cdots,K,\ l = 1,2,3,\cdots,L,\ n = 1,2,3,\cdots,N \quad (6)$$

$$\sum_{l=1}^{L} Q5_{ml} \times Y5_m \leq CPRP_m, \text{ where } m = 1,2,3,\cdots,M,\ l = 1,2,3,\cdots,L \quad (7)$$

$$\left[\sum_{o=1}^{O} Q9_{no} \times Y6_n \cdot Y7_0 + \sum_{p=1}^{P} Q10_{np} \times Y6_n \cdot Y8_p + \sum_{r=1}^{R} Q11_{nr} \times Y6_n \cdot Y9_r + \sum_{i=1}^{I} Q12_{ni} \times Y6_n \cdot Y1_i\right] \leq CPCDR_n,$$
$$\text{where } n = 1,2,3,\cdots,N,\ o = 1,2,3,\cdots,O,\ p = 1,2,3,\cdots,P,\ r = 1,2,3,\cdots,R,\ i = 1,2,3,\cdots,I \quad (8)$$

$$\sum_{n=1}^{N} Q9_{no} \times Y6_n \cdot Y7_0 \leq CPLF_o, \text{ where } n = 1,2,3,\cdots,N,\ o = 1,2,3,\cdots,O \quad (9)$$

$$\sum_{n=1}^{N} Q10_{np} \times Y6_n \cdot Y8_p \leq CPRC_p, \text{ where } n = 1,2,3,\cdots,N,\ p = 1,2,3,\cdots,P \quad (10)$$

$$\sum_{r=1}^{R} Q11_{nr} \times Y6_n \cdot Y9_r \leq CPRF_n, \text{ where } n = 1,2,3,\cdots,N,\ r = 1,2,3,\cdots,R \quad (11)$$

$$\sum_{r=1}^{R} Q13_{ir} \times Y1_i \cdot Y9_r \leq CPRM_i, \text{ where } i = 1,2,3,\cdots,I,\ r = 1,2,3,\cdots,R \quad (12)$$

$$\sum_{r=1}^{R} Q11_{nr} \times Y6_n \cdot Y9_r + \sum_{r=1}^{R} Q13_{ir} \times Y1_i \cdot Y9_r \leq CPRS_r,$$
$$\text{where } i = 1,2,3,\cdots,I,\ n = 1,2,3,\cdots,N,\ r = 1,2,3,\cdots,R \quad (13)$$

$$\sum_{s=1}^{S} Q14_{rs} \times Y9_r \leq CPRS_r, \text{ where } r=1,2,3,\cdots,R,\ s=1,2,3,\cdots,S \tag{14}$$

Balance Constraints (BC):

$$\sum_{i=1}^{I} (Q0_i \times Y1_i) = \sum_{l=1}^{L} D1_l, \text{ where } i=1,2,3,\cdots,I,\ l=1,2,3,\cdots,L \tag{15}$$

$$\sum_{i=1}^{I} (Q0_i \times Y1_i) = \sum_{j=1}^{J} Q1_{ij} \times Y1_i \cdot Y2_j, \text{ where } i=1,2,3,\cdots,I,\ j=1,2,3,\cdots,J \tag{16}$$

$$\sum_{j=1}^{J} Q1_{ij} \times Y1_i \cdot Y2_j = \sum_{j=1}^{J} Q2_{jk} \times Y2_j \cdot Y3_k, \text{ where } i=1,2,3,\cdots,I,\ j=1,2,3,\cdots,J,\ k=1,2,3,\cdots,K \tag{17}$$

$$\sum_{j=1}^{J} Q2_{jk} \times Y2_j \cdot Y3_k = \sum_{l=1}^{L} Q3_{kl} \times Y3_k, \text{ where } j=1,2,3,\cdots,J,\ k=1,2,3,\cdots,K,\ l=1,2,3,\cdots,L \tag{18}$$

$$\sum_{l=1}^{L} Q3_{kl} \times Y3_k \geq \left(\sum_{l=1}^{L} Q4_{lm} \times Y5_m + \sum_{k=1}^{K} Q7_{lk} \times Y3_k + \sum_{n=1}^{N} Q6_{ln} \times Y6_n\right),$$
$$\text{where } k=1,2,3,\cdots,K,\ l=1,2,3,\cdots,L,\ m=1,2,3,\cdots,M,\ n=1,2,3,\cdots,N \tag{19}$$

$$\sum_{l=1}^{L} Q4_{lm} \times Y5_m \leq FCDF_{RR} \times \sum_{l=1}^{L} D1_l, \text{ where } m=1,2,3,\cdots,M,\ l=1,2,3,\cdots,L \tag{20}$$

$$\sum_{l=1}^{L} Q4_{lm} \times Y5_m = \sum_{m=1}^{M} Q5_{ml} \times Y5_m, \text{ where } m=1,2,3,\cdots,M,\ l=1,2,3,\cdots,L \tag{21}$$

$$\left[\sum_{k=1}^{K} Q7_{lk} \times Y3_k + \sum_{n=1}^{N} Q6_{ln} \times Y6_n\right] \leq FCDF_{EOLR} \times \sum_{l=1}^{L} D1_l, \text{ where } k=1,2,3,\cdots,K,\ l=1,2,3,\cdots,L \tag{22}$$

$$\sum_{k=1}^{K} Q7_{lk} \times Y3_k \leq EOLF_{RT} \times \left(FCDF_{EOLR} \times \sum_{l=1}^{L} D1_l\right), \text{ where } k=1,2,3,\cdots,K,\ l=1,2,3,\cdots,L \tag{23}$$

$$\sum_{n=1}^{N} Q6_{ln} \times Y6_n \le EOLF_{CDR} \times \left(FCDF_{EOLR} \times \sum_{l=1}^{L} D1_l \right), \text{ where } n = 1,2,3,\cdots,N,\ l = 1,2,3,\cdots,L \quad (24)$$

$$\sum_{k=1}^{K} Q8_{kn} \times Y3_k \cdot Y6_n + \sum_{l=1}^{L} Q6_{ln} \times Y6_n$$
$$= \left[\sum_{o=1}^{O} Q9_{no} \times Y6_n \cdot Y7_0 + \sum_{p=1}^{P} Q10_{np} \times Y6_n \cdot Y8_p + \sum_{r=1}^{R} Q11_{nr} \times Y6_n \cdot Y9_r + \sum_{i=1}^{I} Q12_{ni} \times Y6_n \cdot Y1_i \right]$$
$$\text{where } k = 1,2,3,\cdots,K,\ l = 1,2,3,\cdots,L,\ n = 1,2,3,\cdots,N,\ o = 1,2,3,\cdots,O,\ p = 1,2,3,\cdots,P,$$
$$r = 1,2,3,\cdots,R,\ i = 1,2,3,\cdots,I \quad (25)$$

$$\sum_{i=1}^{I} Q12_{ni} \times Y6_n \cdot Y1_i = \sum_{i=1}^{I} Q13_{ir} \times Y1_i \cdot Y9_r,$$
$$\text{where } i = 1,2,3,\cdots,I,\ n = 1,2,3,\cdots,N,\ r = 1,2,3,\cdots,R,\ s = 1,2,3,\cdots,S \quad (26)$$

$$\sum_{n=1}^{N} Q11_{nr} \times Y6_n \cdot Y9_r + \sum_{i=1}^{I} Q13_{ir} \times Y1_i \cdot Y9_r = \sum_{s=1}^{S} Q14_{rs} \times Y9_r,$$
$$\text{where } i = 1,2,3,\cdots,I,\ n = 1,2,3,\cdots,N,\ r = 1,2,3,\cdots,R,\ s = 1,2,3,\cdots,S \quad (27)$$

$$\sum_{r=1}^{R} Q14_{rs} \times Y9_r = D2_s, \text{ where } s = 1,2,3,\cdots,S \quad (28)$$

where,

$$Y1_i = 0 \text{ or } 1,\ i = 1,2,3,\cdots,I,\ Y2_j = 0 \text{ or } 1,\ j = 1,2,3,\cdots,J,\ Y3_k = 0 \text{ or } 1, \text{ for } k = 1,2,3,\cdots,K$$

$$Y5_m = 0 \text{ or } 1, \text{ for } m = 1,2,3,\cdots,M,\ Y6_n = 0 \text{ or } 1, \text{ for } n = 1,2,3,\cdots,N,\ Y7_o = 0 \text{ or } 1, \text{ for } o = 1,2,3,\cdots,O$$

$$Y8_p = 0 \text{ or } 1,\ p = 1,2,3,\cdots,P,\ Y9_r = 0 \text{ or } 1, \text{ for } r = 1,2,3,\cdots,R,\ Q0_i \ge 0, \text{ for } i = 1,2,3,\cdots,I$$

$$Q1_{ij} \ge 0, \text{ for } i = 1,2,3,\cdots,I \text{ and } j = 1,2,3,\cdots,J,\ Q2_{jk} \ge 0, \text{ for } j = 1,2,3,\cdots,J \text{ and } k = 1,2,3,\cdots,K$$

$$Q3_{kl} \ge 0, \text{ for } k = 1,2,3,\cdots,K \text{ and } l = 1,2,3,\cdots,L,\ Q4_{lm} \ge 0, \text{ for } l = 1,2,3,\cdots,L \text{ and } m = 1,2,3,\cdots,M$$

$$Q5_{ml} \ge 0,\ m = 1,2,3,\cdots,M \text{ and } l = 1,2,3,\cdots,L,\ Q6_{ln} \ge 0, \text{ for } l = 1,2,3,\cdots,L \text{ and } n = 1,2,3,\cdots,N$$

$$Q7_{lk} \ge 0, \text{ for } l = 1,2,3,\cdots,L \text{ and } k = 1,2,3,\cdots,K,\ Q8_{kn} \ge 0, \text{ for } k = 1,2,3,\cdots,K \text{ and } n = 1,2,3,\cdots,N$$

$$Q9_{no} \geq 0, \text{ for } n = 1,2,3,\cdots,N \text{ and } o = 1,2,3,\cdots,O, \ Q10_{np} \geq 0, \text{ for } n = 1,2,3,\cdots,N \text{ and } p = 1,2,3,\cdots,P$$

$$Q11_{nr} \geq 0, \text{ for } n = 1,2,3,\cdots,N \text{ and } r = 1,2,3,\cdots,R, \ Q12_{ni} \geq 0, \text{ for } n = 1,2,3,\cdots,N \text{ and } i = 1,2,3,\cdots,I$$

$$Q13_{ir} \geq 0, \text{ for } i = 1,2,3,\cdots,I \text{ and } r = 1,2,3,\cdots,R, \ Q14_{rs} \geq 0, \text{ for } r = 1,2,3,\cdots,R \text{ and } s = 1,2,3,\cdots,S$$

Constraints (1) and (2) correspond to the demands of the first customers and second customers respectively. Constraint (3) makes sure that the sum of the outflows from each manufacturer to all the wholesalers does not exceed the capacity of the manufacturers. Constraint (4) makes sure that the sum of the outflows from each manufacturer to all the wholesalers plus sum of the outflows from the each manufacturer to all the resellers minus sum of the outflows from each CDR to all the manufacturers for remanufacturing does not exceed the capacity of the manufacturers. Constraint (5) makes sure that the sum of the outflows from each wholesaler to all the retailers does not exceed the capacity of the wholesalers. Constraint (6) makes sure that the sum of the outflows from each retailer to all the first customers and CDRs does not exceed the capacity of the retailers. Constraint (7) makes sure that the sum of the outflows from each repair center to all the first customers does not exceed the capacity of the repair centers. Constraint (8) makes sure that the sum of the outflows from each CDR to all the remanufacturers, resellers, recyclers and land fillers does not exceed the capacity of the CDRs. Constraint (9) makes sure that the sum of the inflows to land fillers from all CDRs does not exceed the capacity of the land fillers. Constraint (10) makes sure that the sum of the inflows to recyclers from all CDRs does not exceed the capacity of the recyclers. Constraint (11) makes sure that the sum of the outflows from each re-furbisher under CDRs to all the resellers does not exceed the refurbishing capacity of CDRs. Constraint (12) makes sure that the sum of the outflows from each remanufacturer to all the resellers does not exceed the capacity of the remanufacturers. Constraint (13) makes sure that the sum of the inflows to resellers from all the CDRs (refurbished items) and from all the remanufacturers does not exceed the capacity of the resellers. Constraint (14) makes sure that the sum of the outflows from each reseller to all the second customers does not exceed the capacity of the resellers. Constraint (15) makes sure that the sum of the quantities produced by the manufacturers is equal to the sum of the demands of the first customers. Constraint (16) makes sure that the sum of the quantities produced by the manufacturers is equal to the sum of the outflows from the manufacturers. Constraint (17) makes sure that the sum of the inflows to the wholesalers from each manufacturer is equal to the sum of outflows from the wholesalers to the retailers. Constraint (18) makes sure that the sum of the inflows to the retailers from each wholesaler is equal to the sum of the outflows from the retailers to first customers. Constraint (19) makes sure that the sum of the inflows to each first customer from retailer is greater than or equal to the sum of the outflows from each first customers to repair centers and CDRs. Constraint (20) makes sure that the sum of the outflows from each first customer to the repair center is less than or equal to the demand fraction for repair returns of the sum of the demands of the first customers. Constraint (21) makes sure that the sum of the inflows to each repair center from all the first customers is equal to the sum of the outflows from each repair center to all the first customers. Constraint (22) makes sure that the sum of the outflows from each first customer as EOL via retailer or directly to the CDRs is less than or equal to the demand fraction for EOL returns of the sum of the demands of the first customers. Constraint (23) makes sure that the sum of the outflows from each first customer as EOL via retailer to the CDRs is less than or equal to its return fraction of the sum of the EOL returns. Constraint (24) makes sure that the sum of the outflows from each first customer as EOL directly to the CDRs is less than or equal to its return fraction of the sum of the EOL returns. Constraint (25) makes sure that the sum of the EOL returns inflows to each CDR from first customer via retailer and directly is equal to the sum of the flow of CDR fractions exiting from each CDR to all the land fillers, recyclers, resellers and remanufacturers. Constraint (26) makes sure that the sum of the inflows to all the remanufacturers from each CDR is less than or equal to the sum of the outflows from the remanufacturers to resellers. Constraint (27) makes sure that the sum of the inflows to all the resellers from each CDR and each remanufacturer is equal to the sum of the outflows from the resellers to the second customers. Constraint (28) makes sure that the sum of the outflows from each reseller to all the second customers is equal to the sum of the demands of all the second customers.

Thus the mathematical model for the CLSC network design problem formulated as Multi Integer Non-Linear Programming model has been solved by Aravendan and Panneerselvam (2014) [58] using Lingo 14 and the optimized shipment flows of the solved CLSCND model is illustrated in Figure 3. The results of this model are compared with the best hybrid genetic algorithm out the set of four different genetic algorithms proposed in this research.

7. DEVELOPMENT OF HYBRID GENETIC ALGORITHMS

A hybrid genetic algorithm is the fusion or combination of two algorithmic structures i.e. the structure of the main genetic algorithm and a special algorithm specifically incorporated according to the nature of the problem. In this problem, the computation of total cost also includes the transshipment cost of the network in which the main genetic algorithm is fused with the transshipment algorithm which is the combination of VAM and U-V method Algorithms. This hybrid genetic algorithm is also developed with a combination of two selection methods i.e. Elitism and Rank selection. These combinations lead to the formation of variants of HGA developed.

7.1. Preliminaries of Hybrid Genetic Algorithm (HGA)

This section presents the preliminaries of genetic algorithm adopted to develop the variants of .hybrid genetic algorithms.

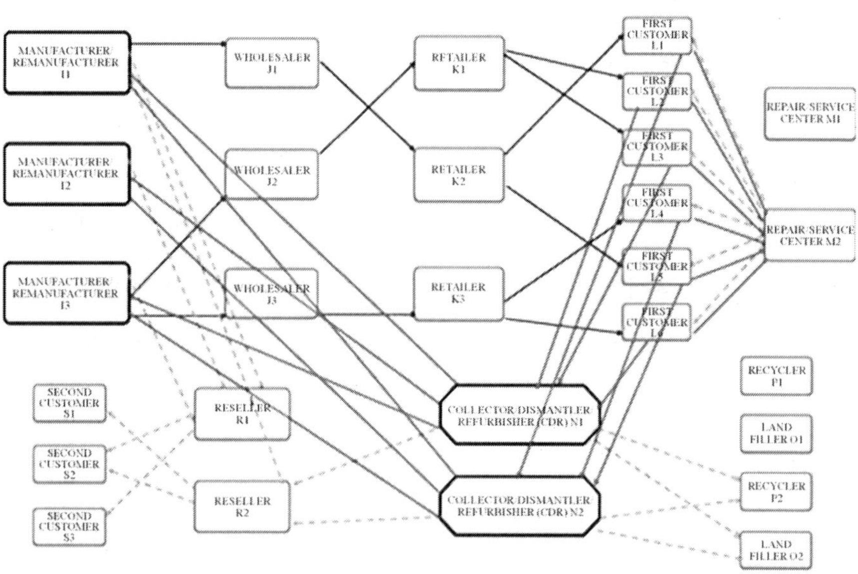

Figure 3. Shipment flows of CLSCND after optimization.

7.1.1. Chromosome Encoding Method

Chromosome Encoding is a process of representing the individual genes in the chromosomes. The representation can be performed using bits, numbers, trees, arrays, lists or any other objects. In this particular problem, binary encoding (using bits) is performed as given as follows.

Binary Encoding

The most common way of encoding is a binary string in which each chromosome is encoded with a binary (bit) string. Each bit in the string represents a gene i.e. node or facility in the CLSC network design. So, every bit string is a solution but not necessarily the best solution. The binary encoding applied in this hybrid genetic algorithm for the example problem is illustrated in Table 1.

7.1.2. Selection Methods

A combination of elitism selection method and rank selection method are applied in all the HGA variants developed. The methods are briefly explained below.

1) Elitism Selection Method

In this method, the first best chromosome or the few best chromosomes are copied to the new population. The rest is done in a classical way. Such individuals can be lost if they are not selected to reproduce or if crossover or mutation destroys them. This method significantly improves the GA's performance.

2) Rank Selection Method

Rank selection method ranks the population and every chromosome receives fitness from the ranking. The worst fitness has the last rank and the best fitness has the first rank. It results with slow convergence but prevents too quick convergence. It also keeps up selection process when the fitness variance is low. It preserves diversity and hence leads to a successful search.

7.1.3. Crossover Methods

An advantage of having more crossover points is that the problem space may be searched more thoroughly. The crossover methods applied for developing the variants of hybrid genetic algorithms are: 1) Two-point crossover method and 2) Uniform crossover method which are discussed as given in Figure 4 and Figure 5 respectively.

7.1.4. Chromosome Replacement Methods

The HGA variants are also developed by varying the chromosome replacement methods applied after crossover and mutation processes. The two methods, viz. both parent replacement method and weak parent replacement method, adopted in this research are briefed as follows.

1) Both Parent Replacement Method

In this method, the fitness values of the off-spring are replaced with that of the parent chromosomes irrespective of the nature of the fitness values. The fitness values of the parent chromosomes are not checked whether they are weaker or stronger than their corresponding offspring.

2) Weak Parent Replacement Method

In this method, the fitness values of the offspring are replaced only if they are stronger than the fitness values of the parent chromosomes. So, the subsequent generations will have chromosomes with better fitness values.

Table 1. Binary encoding of chromosomes in HGA.

Facilities-Nodes/ Chromosomes	M			W			RT			FC						RP		CDR		LF		RC		RM			RS		SC		
	1	2	3	1	2	3	1	2	3	1	2	3	4	5	6	1	2	1	2	1	2	1	2	1	2	3	1	2	1	2	3
Chromosome 1	1	1	1	1	0	1	1	0	1	1	1	1	1	1	1	1	0	1	0	1	0	1	0	1	0	1	1	0	1	1	1
Chromosome 2	1	1	1	0	1	0	0	1	0	1	1	1	1	1	1	0	1	0	1	0	1	0	1	0	1	0	0	1	1	1	1

M—Manufacturer, W—Wholesaler, RT—Retailer, FC—First Customer, RP—Repair Center, CDR—Collector/Dismantler/Re-furbisher, LF—Land Filler, RC—Re-Cycler, RM—Re-Manufacturer, RS—Re-Seller, and SC—Second Customer.

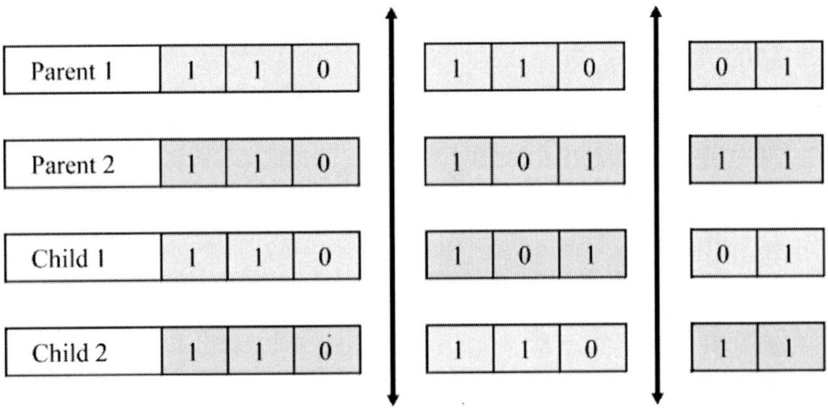

Figure 4. Two-point crossover method. In two-point crossover, two crossover points are chosen and the chromosomes between these points are exchanged between two mated parents. In the above figure, the two thick lines indicate the two crossover points. Thus the contents between these points are exchanged between the parents to produce new children for mating in the next generation.

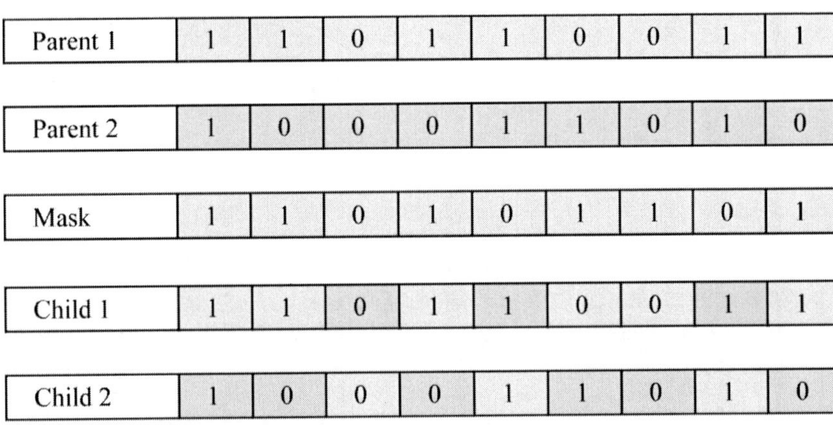

Figure 5. Uniform crossover method. In this method, each gene in the offspring is created by copying the corresponding gene from one or the other parent chosen according to a random generated binary crossover mask of the same length as the chromosomes. Where there is a 1 in the crossover mask, the gene is copied from the first parent, and where there is a 0 in the mask the gene is copied from the second parent. A new crossover mask is randomly generated for each pair of parents. Offspring, therefore, contains a mixture of genes from each parent.

7.2. Steps of Hybrid Genetic Algorithms

The various steps implemented sequentially to develop the hybrid genetic algorithms are furnished as follows.

Step 1: Input the following:

Number of stages = 11 (Manufacturers, Wholesalers, Retailers, First Customers, Repair Centers, Collector/ Dismantler/Re-furbisher, Recycler, Land filler, Remanufacturers, Resellers and Second Customers);

Maximum number of units/nodes in each stage from first to last stage. Ex. 3, 3, 3, 6, 2, 2, 2, 2, 3, 2, 3 (Medium size);

Maximum number of successive populations to be generated (N) = 50;

Maximum number of Chromosomes in Each Population-Population size (L) = 100;

The Generation Count GC = 1.

Step 2: Apply the binary encoding method for the chromosomes to decide on the status (open or close) of the genes (units) in the Chromosomes.

Step 3: Generate a random initial population of n chromosomes (suitable solutions for the problem). Let it be the larger population L = 100.

Step 4: Evaluate the Fitness function f(x) of each chromosome x in the population L. The fitness function is as used in the mathematical proposed by Aravendan and Panneerselvam [58] .

Fitness Function Value FFV = OPC + OPRC + UCC + TC, Where OPC = Opening Cost, OPRC = Operation Cost, UCC = Un-utilized Capacity Cost and TC = Transportation cost. (The Transshipment Algorithm, a combination of VAM and U-V algorithms is incorporated to evaluate the transshipment cost in the Fitness function)

Step 5: Sort the population L by the objective function (fitness function) value in the ascending order for minimization problem.

Step 6: Apply Elite Count (say top 2) and Rank selection; Select a given percentage (say 30%) from the top of the larger population L leaving the elite count, to form a sub-population S.

Step 7: Randomly pick up any two unselected parent chromosomes from the sub-population S. Let them be the parents xand y.

Step 8: Perform the crossover of the parents x and y for a crossover probability P_c (say $P_c = 0.3$) to form their two new offspring x_1 and y_1. If no crossover is performed, then assume the offspring $x_1$and y_1 as the exact copies of the parent chromosomes x and y respectively. (Two point crossover and Uniform crossover methods are applied to form hybrids)

Step 9: Perform the mutation of each of the offspring x_1 and y_1 for a mutation probability P_m (say $P_m = 0.06$).

Step 10: Evaluate the fitness function value of each of the offspring x_1 and y_1.

Step 11: Replace the FFV of parent chromosomes x and y in the larger population L, with the FFV of their respective offspring x_1 and y_1. (Both parent replacement and Weak parent replacement methods are applied to form hybrids)

Step 12: Repeat Step 7 to Step 11 until all the chromosomes in the sub-population S are selected to create offspring.

Step 13: Increase the Generation Count (GC) by 1, i.e. GC = GC + 1.

Step 14: If GC ≤ N, then go to Step 5, else go to Step 15.

Step 15: Identify the chromosome in the larger population L, which has the best fitness function valueand printthe corresponding results.

Step 16: Stop.

7.3. Development of Variants of Hybrid Genetic Algorithms

The variants of hybrid genetic algorithm are developed by varying the GA parameters like selection method, crossover method and replacement method. Four hybrid genetic algorithms are thus constructed, viz. HGA1, HGA2, HGA3 and HGA4 with the following combination.

HGA1 has the combination of 2 point crossover method and both parent replacement method.

HGA2 has the combination of 2 point crossover method and weak parent replacement method.

HGA3 has the combination of uniform crossover method and both parent replacement method.

HGA4 has the combination of uniform crossover method and weak parent replacement method.

The configurations of these four hybrid genetic algorithms are illustrated in Figure 6.

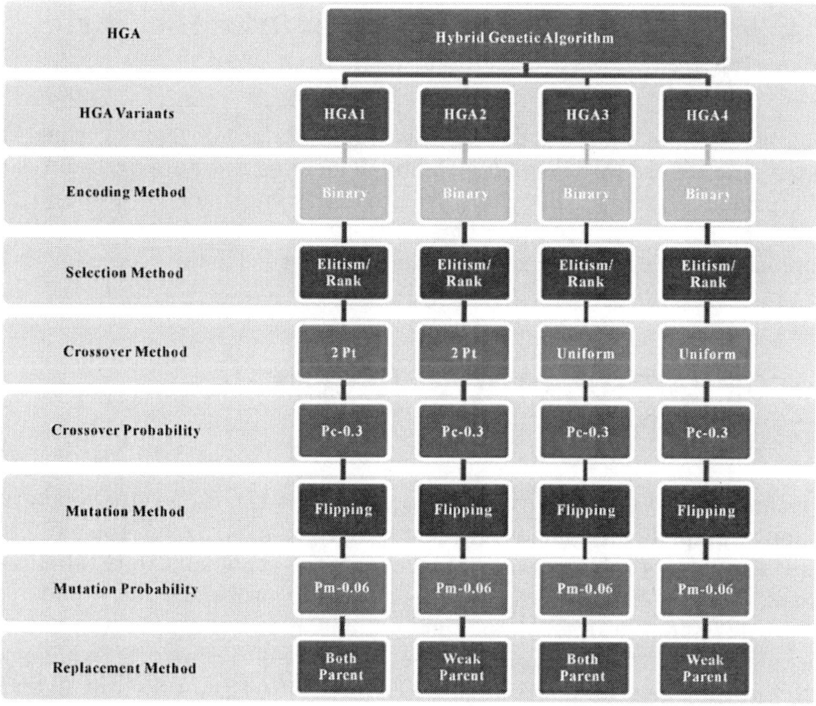

Figure 6. Configuration of the four variants of hybrid genetic algorithm.

Table 2. Experiments/replications with respect to problem sizes.

Expt. No./ Problem Size	No. of Repns./ Expt.	No. of Nodes at the Various Stages in the Closed Loop Supply Chain Network Design										
		M	W	RT	FC	RP	**CDR**	LF	RC	RM	RS	SC
1 Small	6	2	2	2	4	2	2	2	2	2	2	2
2 Medium	6	3	3	3	6	2	2	2	2	3	2	3
3 Large	6	4	4	4	8	2	2	2	2	4	2	4

M—Manufacturer, W—Wholesaler, RT—Retailer, FC—First Customer, RP—Repair Center, CDR—Collector/Dismantler/Re-furbisher, LF—Land Filler, RC—Re-Cycler, RM—Re-Manufacturer, RS—Re-Seller, and SC—Second Customer.

7.4. Development of Source Codes for Hybrid Genetic Algorithms

The programme source codes for the four HGAs are developed using C#. Input data are fed by developing and implementing the store procedures in Microsoft SQL server. The whole programme is implemented in the Visual Basic. NET ver. 4.5 platform. This strategy has remarkably reduced the running time of the hybrid genetic algorithm for solving the problem.

8. COMPARISON OF HYBRID GENETIC ALGORITHMS

The four hybrid genetic algorithms presented in this paper are compared using a complete factorial experiment with two factors, viz. "Problem Size" and "Algorithms". The number of levels for the problem size is three, viz. small, medium and large. In all the three size problems, the number of nodes at manufacturer, wholesaler, retailer, first customers and second customers are varied according to the size of the problems. The number of levels for the algorithm is 4, viz. HGA1, HGA2, HGA3 and HGA4. In all these four algorithms, the selection method applied is common i.e. the combination of Elitism and Rank methods, but the crossover methods and chromosome replacement methods are different as illustrated in Figure 6 of previous section. The number of replications under each experimental combination is 6. The descriptions of the problems are shown in Table 2 and those of the algorithms are shown in Table 3.

Table 3. Experiments/replications with respect to HGA parameters.

Expt. No./ Problem Size	Selection Method	Crossover Method		Crossover Probability	Mutation Probability	Parent Chromosome Replacement	
1 Small	Rank	2 pt.	Uniform	0.3	0.06	Both Parent	Weak Parent
2 Medium	Rank	2 pt.	Uniform	0.3	0.06	Both Parent	Weak Parent
3 Large	Rank	2 pt.	Uniform	0.3	0.06	Both Parent	Weak Parent

The replications of the experiment were carried out in the above combinations for all the three problem sizes vis-à-vis the four HGAs and the results are plotted in Table 4.

The different hypotheses proposed are as listed below.

- Whether there are significant differences among the problem sizes in terms of the total cost;
- Whether there are significance differences among the algorithms in terms of the total cost;
- Whether there are significance differences among the interaction terms of problem size and algorithm in terms of the cost.

The results of ANOVA for the data given in Table 4 are shown in Table 5. From Table 5, it is clear that there are significant differences among the algorithms, HGA1, HGA2, HGA3 and HGA4, because the corresponding p value is less than 0.05 (significance level).

Table 4. Experiment results for HGAs.

Expt. No./ Problem Size	Replications- Transportation Costs	Hybrid Genetic Algorithms (HGA)-Variants			
		HGA1	HGA2	HGA3	HGA4
1 Small	Repln.1: TC = 1	118209336	118208336	118209336	115020093
	Repln.2: TC = 3	116394408	116394408	119150408	115057808
	Repln.3: TC = 5	122613480	121389416	122613480	121253416
	Repln.4: TC = 6	122681016	121479480	122681016	121264802
	Repln.5: TC = 8	122816088	121916088	122806088	121716288
	Repln.6: TC = 10	122951160	122951160	122951150	121932160
2 Medium	Repln.1: TC = 1	165602183	162519783	166398115	155925683
	Repln.2: TC = 3	166398115	162788515	168316549	158710580
	Repln.3: TC = 5	168316549	168316549	168669898	159978749
	Repln.4: TC = 6	168669898	168669898	169105000	160969678
	Repln.5: TC = 8	166278430	169105000	169699983	162614783
	Repln.6: TC = 10	173616784	173616784	182700784	164954088
3 Large	Repln.1: TC = 1	325416926	324416826	325416926	325406921
	Repln.2: TC = 3	325766378	325756378	325776378	325696378
	Repln.3: TC = 5	326125830	326015830	326115930	326015830
	Repln.4: TC = 6	326773600	326582800	327133800	326637800
	Repln.5: TC = 8	327025008	326925008	327025008	326926008
	Repln.6: TC = 10	327230260	326989460	327230260	327230260

Hence, the best algorithm is obtained using Duncan's multiple range tests. The result of Duncan's multiple range tests are shown in Figure 7. From this figure, it is clear that the algorithm HGA4 is significantly different from other algorithms and its total cost is the least among the total costs all the algorithms. Hence, the hybrid genetic algorithm HGA4 is proved the best among all the four proposed hybrid genetic algorithms.

Table 5. Results of ANOVA for comparison of algorithms.

Source	Type III Sum of Squares	Degree of Freedom df	Mean Square	F	Sig	Remarks
Algorithm	1.677E+14	3	5.591E+13	6.417	0.001	Significant
Problem size	5.596E+17	2	2.798E+17	32,114.712	0.000	Significant
Algorithm* Problem Size	1.931E+14	6	3.218E+13	3.694	0.003	Significant
Error	5.227E+14	60	8.712E+12			
Total	5.605E+17	71				

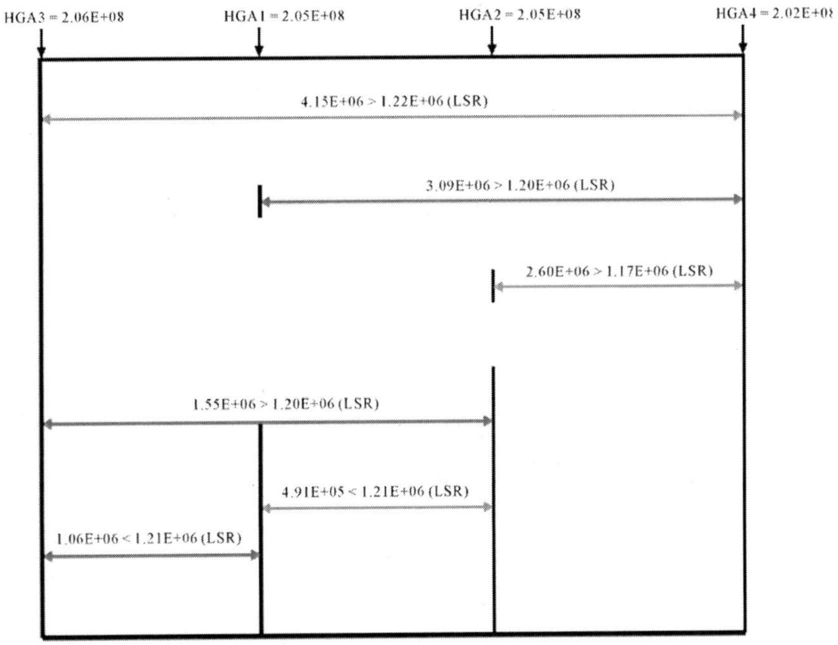

Figure 7. Results of Duncan's multiple range test.

9. COMPARISON OF BEST HYBRID GENETIC ALGORITHM WITH MATHEMATICAL MODEL

In the previous section, it is proved that the hybrid genetic algorithm 4 (HGA4) is proved to be the best among all the four proposed algorithms. Hence, in this section, it is benchmarked against the results of the mathematical model presented in section 6 using a complete factorial experiment in which the problem size is considered as one factor and the method of solving a problem is considered as another factor. The number of levels for the problem size is 3, viz. small, medium and large and the number of levels for the method is 2, viz. HGA4 and Model. The number of replications carried out under each experimental combination is 6. The results as per this complete factorial experiment are shown in Table 6. It is seen that the model did not give results even after long time execution for large size problems. Hence, the number of levels of problem size is reduced to 2, viz. small and medium.

The different hypotheses of this comparison are listed below.

- Whether there are significant differences among the problem sizes in terms of the total cost;
- Whether there are significant differences among the two methods in terms of the total cost;
- Whether there are significant differences among the interaction terms of problem size and method in terms of the cost.

Table 6. Experimental results of HGA4 and model.

Expt. No./ Problem Size	Replications- Transportation Costs	HGA4	Mathematical Model
1 Small	Repln.1: TC = 1	115020093	114810000
	Repln.2: TC = 3	115057808	114946700
	Repln.3: TC = 5	121253416	121066100
	Repln.4: TC = 6	121264802	121156700
	Repln.5: TC = 8	121716288	121302200
	Repln.6: TC = 10	121932160	121459500
2 Medium	Repln.1: TC = 1	155925683	155849900
	Repln.2: TC = 3	158710580	157710500
	Repln.3: TC = 5	159978749	159578800
	Repln.4: TC = 6	160969678	160503100
	Repln.5: TC = 8	162614783	162364900
	Repln.6: TC = 10	164954088	164234900
3 Large	Repln.1: TC = 1	325406921	No Solution
	Repln.2: TC = 3	325696378	No Solution
	Repln.3: TC = 5	326015830	No Solution
	Repln.4: TC = 6	326637800	No Solution
	Repln.5: TC = 8	326926008	No Solution
	Repln.6: TC = 10	327230260	No Solution

The results of ANOVA executed for the experimental results shown in Table 6 are shown in Table 7. From the results shown in Table 6, it is clear that there is no significant difference between the levels of the factor "method", which means that the there is no significant difference between the results of HGA4 and the model. So, HGA4 can be equated to model in terms of given solution for small size and medium size problems.

Table 7. Results of ANOVA for the comparison of HGA4 with model.

Source	Type III Sum of Squares	Degree of Freedom df	Mean Square	F	Sig.	Remarks
Algorithm	8.121E+11	1	8.121E+11	0.079	0.782	Not Significant
Problem size	1.010E+16	1	1.010E+16	979.781	0.000	Significant
Algorithm* Problem Size	8.261E+10	1	8.261E+10	0.008	0.930	Not Significant
Error	2.0627E+14	20	1.031E+13			
Total	1.031E+16	23				

10. CONCLUSIONS

In this paper, four different hybrid genetic algorithms have been proposed and they are compared through a complete factorial experiment with two factors, viz. problem size (three problem sizes) and Algorithm (four algorithms). It is found that there are significant differences among the four algorithms, viz. HGA1, HGA2, HGA3 and HGA4. Hence, using Duncan's Multiple Range Test, it is found that the algorithm HGA4 is the best. In the next stage, HGA4 is compared with model using a complete factorial experiment with two factors, viz. Problem Size (Two sizes) and Method (HGA4 and Model). It is found that there is no significant difference between HGA4 and Model. So, HGA4 is proved to be superior in terms of giving very near optimal solution for small and medium size problems and provides the best solution when compared to all other hybrid genetic algorithms, viz. HGA1, HGA2 and HGA3.

Scope and Future Research Directions

The model and the hybrid genetic algorithm developed have good scope in the industrial application as they can be applied to a wide range of industries like automobiles, consumer electronics, textile apparels, fashion leather garments, luxury goods, footwear, home appliances, industrial equipment and machines, etc. As the incorporation of hybrid centers/facilities at the same location in the forward and reverse chains eliminates the establishment and maintenance costs of the separate facilities at different locations, the overall cost of the closed loop supply chain is minimized remarkably. The model and algorithm are proposed for the single product and single period system. Further research on the network design for multi-products and multi-periods CLSC system would be beneficial to the business sector. The future direction of research could also be an extension of this research work towards developing CLSC network design with incapacitated facilities to deal with the stochastic/uncertain demands of the customers.

ACKNOWLEDGEMENTS

The authors profusely thank M/S. Lindo Systems Inc., Chicago, USA for their technical support by providing the license of the extended version of LINGO 15 optimization software package. The authors also thank the anonymous referees for their constructive comments and suggestions, which have been incorporated in this research paper.

CITE THIS PAPER

MuthusamyAravendan,RamasamyPanneerselvam, (2015) Development and Comparison of Hybrid Genetic Algorithms for Network Design Problem in Closed Loop Supply Chain. *Intelligent Information Management*,**07**,313-338. doi: 10.4236/iim.2015.76025

REFERENCES

1. Goldberg, D.E. and Deb, K. (1991) A Comparative Analysis of Selection Schemes Used in Genetic Algorithms. In: Rawlins, G., Ed., Foundations of Genetic Algorithms, Morgan Kaufmann Publishers, San Francisco, 69-93.

2. Srinivas, N. and Deb, K. (1994) Multi-Objective Optimization Using Non-Dominated Sorting in Genetic Algorithms. Journal of Evolutionary Computation, 2, 221-248.

3. Golub, M. (1996) An Implementation of Binary and Floating Point Chromosome Representation in Genetic Algorithm. Proceedings of

the 18th International Conference ITI, Pula, 18-21 June 1996, 417-422.

4. Deb, K. (1999) An Introduction to Genetic Algorithm. Sadhana, 24, 293-315.
5. Lin, W.-Y., Lee, W.-Y. and Hong, T.-P. (2003) Adapting Crossover and Mutation Rates in Genetic Algorithms. Journal of Information Science and Engineering, 19, 889-903.
6. Sim, E., Jung, S., Kim, H. and Park, J. (2004) A Generic Network Design for a Closed-Loop Supply Chain Using Genetic Algorithm. Lecture Notes in Computer Science, 3103, 1214-1225.
7. Yeh, W.-C. (2005) A Hybrid Heuristic Algorithm for the Multistage Supply Chain Network Problem. International Journal of Advanced Manufacturing Technology, 26, 675-685.
8. Altiparmak, F., Gen, M., Lin, L. and Paksoy, T. (2006) A Genetic Algorithm Approach for Multi-Objective Optimization of Supply Chain Networks. Computers & Industrial Engineering, 51, 196-215.
9. Min, H., Ko, C.S. and Ko, H.J. (2006) The Spatial and Temporal Consolidation of Returned Products in a Closed-Loop Supply Chain Network. Computers & Industrial Engineering, 51, 309-320.
10. El-Mihoub, T.A., Hopgood, A.A., Nolle, L. and Battersby, A. (2006) Hybrid Genetic Algorithms: A Review. Engineering Letters, 13, 124-137.
11. Zhou, G., Cao, Z., Qi, F. and Cao, J. (2006) A Genetic Algorithm Approach on a Logistics Distribution System with Uncertain Demand and Product Return. World Journal of Modelling and Simulation, 2, 99-108.
12. Schultmann, F., Zumkeller, M. and Rentz, O. (2006) Modeling Reverse Logistic Tasks within Closed-Loop Supply Chains: An Example from the Automotive Industry. European Journal of Operational Research, 171, 1033-1050.
13. Altiparmak, F., Gen, M., Lin, L. and Karaoglan, I. (2007) A Steady-State Genetic Algorithm for Multi-Product Supply Chain Network Design. Computers & Industrial Engineering, 56, 521-537.
14. Lin, L., Gen, M. and Wang, X. (2007) A Hybrid Genetic Algorithm for Logistics Network Design with Flexible Multistage Model. International Journal of Information Systems for Logistics and Management, 3, 1-12.

15. Staikos, T. and Rahimifard, S. (2007) A Decision-Making Model for Waste Management in the Footwear Industry. Journal of Production Research, 45, 4403-4422.

16. Ko, H.J. and Evans, G.W. (2007) A Genetic Algorithm-Based Heuristic for the Dynamic Integrated Forward/Reverse Logistics Network for 3PLs. Computers & Operations Research, 34, 346-366.

17. Min, H. and Ko, H.-J. (2008) The Dynamic Design of a Reverse Logistics Network from the Perspective of Third-Party Logistics Service Providers. International Journal of Production Economics, 113, 176-192.

18. Belgasmi, N., Said, L.B. and Ghedira, K. (2008) Genetic Optimization of the Multi-Location Transshipment Problem with Limited Storage Capacity. Proceedings of the 18th European Conference on Artificial Intelligence, Patras, 21-25 July 2008, 563-567.

19. Farahania, R.Z. and Elahipanaha, M. (2008) A Genetic Algorithm to Optimize the Total Cost and Service Level for Just-in-Time Distribution in a Supply Chain. International Journal of Production Economics, 111, 229-243.

20. Lee, D.H. and Dong, M. (2008) A Heuristic Approach to Logistics Network Design for End-of-Lease Computer Products Recovery. Transportation Research Part E, 44, 455-474.

21. Yun, Y.S., Moon, C. and Kim, D. (2009) Hybrid Genetic Algorithm with Adaptive Local Search Scheme for Solving Multi-Stage Based Supply Chain Problems. Computers & Industrial Engineering, 56, 821-838.

22. Sourirajan, K., Ozsen, L. and Uzsoy, R. (2009) A Genetic Algorithm for a Single Product Network Design Model with Lead Time and Safety Stock Considerations. European Journal of Operational Research, 197, 599-608.

23. Lee, J.-E., Gen, M. and Rhee, K.-G. (2009) Network Model and Optimization of Reverse Logistics by Hybrid Genetic Algorithm. Computers & Industrial Engineering, 56, 951-964.

24. Gen, M., Lin, L. and Jo, J.-B. (2009) Evolutionary Network Design by Multi Objective Hybrid Genetic Algorithm. Intelligent and Evolutionary Systems, SCI 187, 105-121.

25. Lin, L., Gen, M. and Wang, X. (2009) Integrated Multistage Logistics Network Design by Using Hybrid Evolutionary Algorithm. Computers & Industrial Engineering, 56, 854-873.

26. Salema, M.I.G., Povoa, A.P.B. and Novais, A.Q. (2009) A Strategic and Tactical Model for Closed-Loop Supply Chains. OR Spectrum, 31, 573-599.

27. Costa, A., Celano, G., Fichera, S. and Trovato, E. (2010) A New Efficient Encoding/Decoding Procedure for the Design of a Supply Chain Network with Genetic Algorithms. Computers & Industrial Engineering, 59, 986-999.

28. Pishvaee, M.S., et al. (2010) A Memetic Algorithm for Bi-Objective Integrated Forward/Reverse Logistics Network Design. Computers & Operations Research, 37, 1100-1112.

29. Zarei, M., Mansour, S., Kashan, A.H. and Karimi, B. (2010) Designing a Reverse Logistics Network for End-of-Life Vehicles Recovery. Mathematical Problems in Engineering, 2010, 1-16.

30. Kannan, G., Sasikumar, P. and Devika, K. (2010) A Genetic Algorithm Approach for Solving a Closed Loop Supply Chain Model: A Case of Battery Recycling. Applied Mathematical Modeling, 34, 655-670.

31. El-Sayed, M., Afia, N. and El-Kharbotly, A. (2010) A Stochastic Model for Forward-Reverse Logistics Network Design under Risk. Computers & Industrial Engineering, 58, 423-431.

32. Kaya, Y., Uyar, M. and Tekdn, R. (2011) A Novel Crossover Operator for Genetic Algorithms: Ring Crossover.

33. Khajavi, L.T., Seyed-Hosseini, S.M. and Ahmad Makui, A. (2011) An Integrated Forward/Reverse Logistics Network Optimization Model for Multi-Stage Capacitated Supply Chain. iBusiness, 3, 229-235.

34. Nandita, A. (2011) The Apparel Aftermarket in India—A Case Study Focusing on Reverse Logistics. Journal of Fashion Marketing and Management, 15, 211-227.

35. Hosseinzadeh, M. and Roghanian, E. (2012) An Optimization Model for Reverse Logistics Network under Stochastic Environment Using Genetic Algorithm. International Journal of Business and Social Science, 3, 249-264.

36. Kumar, R. and Jyotishree, (2012) Novel Encoding Scheme in Genetic Algorithms for Better Fitness. International Journal of Engineering and Advanced Technology (IJEAT), 1, 214-218.

37. Bozorgirad, S., Desa, M.I. and Wibowo, A. (2012) Genetic Algorithm Enhancement to Solve Multi Source Multi Product Flexible Multistage Logistics Network. IJCSI International Journal of Computer Science Issues, 9, 157-164.

38. Mehdizadeha, E. and Afrabandpeia, F. (2012) Design of a Mathematical Model for Logistic Network in a Multi-Stage Multi-Product Supply Chain Network and Developing a Meta Heuristic Algorithm. Journal of Optimization in Industrial Engineering, 10, 35-43.

39. Iris, C. and Serdarasan, S. (2012) A Review of Genetic Algorithm Applications in Supply Chain Network Design. In: Kahraman, C., Ed., Computational Intelligence Systems in Industrial Engineering, Atlantis Press Book, Paris, 209-236.

40. Zaki, S.A., Mousa, A.A.A., Geneedi, H.M. and Elmekawy, A.Y. (2012) Efficient Multi-Objective Genetic Algorithm for Solving Transportation, Assignment and Transshipment Problems. Applied Mathematics, 3, 92-99.

41. Ozkir, V. and Basligil, H. (2012) Modeling Product-Recovery Processes in Closed-Loop Supply-Chain Network Design. International Journal of Production Research, 50, 2218-2233.

42. Rafsanjani, M.K. and Eskandari, S. (2013) Using Segment Based Genetic Algorithm with Local Search to Find Approximate Solution for Multi-Stage Supply Chain Network Design Problem. Cankaya University Journal of Science and Engineering, 10, 185-201.

43. Lee, J.-E., Chung, K.-Y., Lee, K.-D. and Gen, M. (2015) A Multi-Objective Hybrid Genetic Algorithm to Minimize the Total Cost and Delivery Tardiness in a Reverse Logistics. Multimedia Tools and Applications, 74, 9067-9085.

44. Soleimani, H., Esfahani, M.H. and Shirazi, M.A. (2013) Designing and Planning a Multi Period Multi Product Closed-Loop Supply Chain Utilizing Genetic Algorithm. International Journal of Advanced Manufacturing Technology, 68, 917-931.

45. Lee, J.-E. and Lee, K.-D. (2013) Modeling and Optimization of Closed-Loop Supply Chain Considering Order or Next Arrival of

Goods. International Journal of Innovative Computing, Information and Control, 9, 3639-3654.

46. Teodoro, F.G.S., Lima, C.A.M. and Peres, S.M. (2013) Supply Chain Management and Genetic Algorithm: Introducing a New Hybrid Genetic Crossover Operator.

47. Ramezani, M., Bashiri, M. and Moghaddam, R.T. (2013) A New Multi-Objective Stochastic Model for a Forward/Reverse Logistic Network Design with Responsiveness and Quality Level. Applied Mathematical Modeling, 37, 328-344.

48. Rosa, V.D., Gebhard, M., Hartmann, E. and Wollenweber, J. (2013) Robust Sustainable Bi-Directional Logistics Network Design under Uncertainty. International Journal of Production Economics, 145, 184-198.

49. Mahmoudi, H., Fazlollahtabar, H. and Mahdavi, I. (2013) Mathematical Modeling for Minimizing Costs in a Multilayer Multi-Product Reverse Supply Chain. Industrial Engineering Management, 2, 1-6.

50. Hafeti, S.M. and Jolai, F. (2013) Robust and Reliable Forward-Reverse Logistics Network Design under Demand Uncertainty and Facility Disruptions. Applied Mathematical Modeling, 38, 2630-2647.

51. Cardoso, S.R., Paula, F.D.A., Povoa, B. and Relvas, S. (2013) Design and Planning of Supply Chains with Integration of Reverse Logistics Activities under Demand Uncertainty. European Journal of Operational Research, 226, 436-451.

52. Devika, K., Jafarian, A. and Nourbakhash, V. (2014) Designing a Sustainable Closed-Loop Supply Chain Network Based on Triple Bottom Line Approach: A Comparison of Metaheuristics Hybridization Techniques. European Journal of Operational Research, 235, 594-615.

53. Asghari, M. and Nezhadali, S. (2014) A Non-Dominated Sorting Genetic Algorithm for Sustainable Reverse Logistics Network Design. Proceedings of the 2014 International Conference on Industrial Engineering and Operations Management, Bali, 7-9 January 2014, 2426-2436.

54. Aggarwal, S., Garg, R. and Goswami, P. (2014) A Review Paper on Different Encoding Schemes Used in Genetic Algorithms.

International Journal of Advanced Research in Computer Science and Engineering, 4, 596-600.

55. Demirel, N., Ozceylan, E., Paksoy, T. and Gokcen, H. (2014) A Genetic Algorithm Approach for Optimizing a Closed-Loop Supply Chain Network with Crisp and Fuzzy Objectives. International Journal of Production Research, 52, 3637-3664.

56. Dzupire, N.C. and Gyekye, Y.N. (2014) A Multi-Stage Supply Chain Network Optimization Using Genetic Algorithms. Mathematical Theory and Modeling, 4, 18-28.

57. Rad, S.Y.B., Desa, M.I. and Azari, S.D. (2014) Model and Solve the Bi-Criteria Multi Source Flexible Multistage Logistics Network. International Journal of Advanced Computer Science and Information Technology (IJACSIT), 3, 50-69.

58. Aravendan, M. and Panneerselvam, R. (2014) An Integrated Multi-Echelon Model for a Sustainable Closed Loop Supply Chain Network Design. Intelligent Information Management, 6, 257-279.

59. Sarrafhaa, K., Rahmatia, S.H.A., Niaki, S.T.A. and Talab, A.Z. (2015) A Bi-Objective Integrated Procurement, Production, and Distribution Problem of a Multi-Echelon Supply Chain Network Design: A New Tuned MOEA. Computers & Operations Research, 54, 35-51.

9

CHAPTER

COMPARATIVE STUDY OF DIFFERENT REPRESENTATIONS IN GENETIC ALGORITHMS FOR JOB SHOP SCHEDULING PROBLEM

Vedavyasrao Jorapur*, V. S. Puranik, A. S. Deshpande, M. R. Sharma

Visvesvaraya Technological University, Belgaum, India

ABSTRACT

Due to NP-Hard nature of the Job Shop Scheduling Problems (JSP), exact methods fail to provide the optimal solutions in quite reasonable computational time. Due to this nature of the problem, so many heuristics and meta-heuristics have been proposed in the past to get optimal or near-optimal solutions for easy to tough JSP instances in lesser computational time compared to exact methods. One of such heuristics is genetic algorithm (GA). Representations in GA will have a direct impact on computational time it takes in providing optimal or near optimal solutions. Different representation schemes are possible in case of Job Scheduling Problems. These schemes in turn will have a higher impact on the performance of GA. It is intended to show through this paper, how these representations will perform, by a comparative analysis based on average deviation, evolution of solution over entire generations etc.

Keywords: Job Shop Scheduling, Genetic Algorithm, Genetic Representation, Conceptual Model

1. INTRODUCTION

Scheduling is a decision-making process which deals with allocation of resources to tasks over given time-periods and its goal is to optimize one or more objective functions. A scheduling problem is represented by triplet $\alpha/\beta/\gamma$. α field describes machine environment; β field provides details of processing characteristics and constraints and γ field describes the objective function to be minimized. Being essentially a combinatorial optimization problem, job shop scheduling has caught the attention of researchers in the last so many years for optimized performance. Combinatorial optimization problems can be classified as easy and hard. Problems which are polynomialy solvable with limited number of variables are treated easy and are called P. The notion polynomial solvable depends on the type of encoding. It is assumed that problems describing numerical data are binary encoded and the number of steps involved in solving these increases exponentially with increase in length of string and hence computational time will be enormously large and treated to be hard problems. Job scheduling problems belong to this category and are termed NP-Hard [1] . In the practical manufacturing environment, the scale of job shops is generally much larger than that of JSSP bench mark instances considered in theoretical research. Optimization algorithms for job shop scheduling usually proceed by Branch and Bound and among the most recent and successful, ones are those of Carlier and Pinson (1989) and Applegate and Cook (1991) [2] . Approximation procedures or heuristics were initially developed on the basis of priority rules or dispatching rules. The quality of solutions generated by these procedures leave plenty of room for improvement (1998) [3] . Therefore, traditional or meta-heuristic algorithms can hardly be able to solve such problems satisfactorily. In manufacturing workshops, availability of computational resources is much less than the laboratories which lead to difficulty in exploring all possible feasible solutions. Under such circumstances, it is reasonable to reduce the search space and range to only promising areas. The very idea of using constructive heuristics and heuristic search algorithms for larger problem sizes of JSSP is the computational expensive nature of enumerative techniques and Lagranngian algorithms. According to Osman (1996), a heuristic search "is an iterative generation process which guides a subordinate heuristic by combining intelligently different concepts for exploring and exploiting the search spaces".

Extensive use of genetic algorithms to solve job shop scheduling problems can be seen through literature survey [4] . However, how effectively genetic algorithms can be used in JSSP case is not completely explored. In this context, some direction is provided by Tamer F. Abdelmaguid [5] in his paper. Ga's are based on an abstract model of natural evolution, such that quality of individuals builds to the highest level compatible with the environment (constraints of the problem). (Holland, 1975; Goldberg, 1989)

Representations in GA environment applied so far in job shop scheduling can be classified into nine categories as given by Cheng et al. (1996):

1) Operation based 2) Job based 3) Job pair relation based 4) Completion time based 5) Random keys 6) Preference list based 7) Priority rule based 8) Disjunctive graph based. 9) Machine based.

Nine categories mentioned above can be grouped into two basic encoding approaches—direct and indirect encoding. In direct approach, a Π_j schedule is encoded as a chromosome and genetic operators are used to evolve better individual ones. Categories 1 to 5 are examples of this category. In case of indirect approach, a sequence of decision preferences will be encoded into a chromosome. In this, encoding, genetic operators are applied to improve the ordering of various preferences and a Π_j schedule is then generated from the sequence of preferences. Categories 6 to 9 are examples of this category [6] . These representations need to be studied in case of job shop scheduling problems to compare their performance criteria to generate optimal or near optimal solutions, even though computational comparison of different representations is reported in a tutorial paper by Cheng, Gen and Tsujimura [6] . A report by Anderson, Glass and Potts [7] , conducted with different metaheuristics approaches including four different GA implementations, lacks in consistency as well as coherence as regards number of test problems being tested with requisite number of runs.

The rest of this paper is organized as follows: We will start with mathematical models with certain assumptions that have been used in next section followed by the literature review on the different GA representations used in the case of JSP. Followed by review of GA representations, we will discuss regarding different GA operators frequently used by researchers and our own views on adding other operators not discussed so far. Now, we will analyze the experimental results conducted followed by the conclusion provided in the final part of this paper.

2. PROBLEM FORMULATION

Since it is an important practical problem, some authors have formulated various JSP models based on different production situations and problem assumptions. The most common assumptions in case of JSP are:

1) A machine may process more than one job at a time;
2) No job may be processed by more than one machine at a time;
3) The sequence of machines which a job visits is completely specified and has a linear precedence structure;
4) Processing times are known. All the processing times are assumed to be integers;
5) Each job must be processed on each machine only once. There is no recirculation;
6) Set-up times are assumed zero;
7) Pre-emption is not allowed.

Let "J" represent a set of jobs and each job will be processed on a set of machines in a particular order. Let I = (1.....v) represent the operation indexes. The operation indexes are assigned such that for a job $k \in \mathrm{J}$, the subset of consecutive indexes $\mathrm{I}_k = \langle \beta_k, \beta_k + 1, \beta_k + 2, \cdots \omega_k \rangle \subseteq \mathrm{I}$, is a subset containing indexes for that job. Now from the subset I_k depending on the priority operation with higher or lower value is processed first. Let p_i be the processing time of ith operation, the job which it belongs to is j(i) and the machine on which ith operation carried is m(i).

Now the objective of scheduling process is to determine the start time st_i of an operation $i \in \mathrm{I}$. While assigning a job to a machine based on above calculations following constraints should be taken into consideration viz. The technological constraints will take care of order of operations to be carried out on a job and a second set of constraint will take care of conflict of two jobs to be processed on the same machine simultaneously. Accordingly:

$st_i + p_i \leq st_{i+1}$.. Is the equation to satisfy technological constraints (1)

and

$$st_i \geq st_j + p_j \text{ Or } st_j \geq st_i + p_i$$

(2)

Is the equation to satisfy the conflict of two jobs on the same machine at the same time.

$$\forall i, j \in \mathrm{I} \text{ where } m(i) = m(j) \text{ and } j(i) \neq j(j)$$

Different total cost functions that can be studied are

$F_{max.}(C) := max.\{f_i(C_i) | i = 1, \cdots, n\}$ ...Is called Bottleneck objective and

$\sum f_i(C) = \sum_{i=1}^{i=n} f_i(C_i)$ ... Is called Sum Objective.

The most common objective functions are the make span max $\{(C_i) | i = 1, \cdots, n\}$ and total flow time $\sum_{i=1}^{n}(C_i)$, and weighted (total) flow time $\sum_{i=1}^{n} w_i \cdot C_i$. We have considered the minimization of make span as our objective function. Mannes' [8] proposed an integer linear programming model (ILP) which uses different forms of binary variables. This model has gained larger interest in the research community due to small number of variables considered in the model. The technological constraints of Equation (1) are analogous to a series of consecutive activities that are carried out in project scheduling. This analogy has motivated importing project networks into JSP environment. To represent disjunctive constraints as in Equation (2), additional sets of arcs required. This is achieved in a disjunctive graph model [9] and PIAN model [10] .

In the disjunctive graph model, a disjunctive arc is defined between a pair of operations that share the machine. Each disjunctive arc is assigned a binary decision variable such that selection on the value that variable defines the length and direction of each disjunctive arc. This is to the Mannes' model. Very efficient algorithms like immediate selections and shifting bottleneck heuristics were proposed by Carlier [11] and Adams [12] and Lars Monch [13] , which are derived from disjunctive graph model.

A variable notation of the type $\mathbf{X}_{i,t}^{m}$ = 1...if operation 'i' is processed on machine 'm' in unit time 't' = 0...otherwise.

In ILP model was proposed by Bowman [14] . Wagner [15] proposed a model where a variable notation of the type: $X_{i,l}^{m}$ = 1…if operation 'i' takes 'i^th^' position in the processing sequence on machine 'm' = 0...otherwise.

And. Mannes' [8] proposed a model where a variable notation of the type: $X_{i,j}^{m}$ = 1…if operation 'i' is processed prior to operation 'j' on machine m. = 0...otherwise.

3. REPRESENTATION OF THE PROBLEM IN GA AND GA OPERATORS

Darwin's principle "survival of the fittest" can be used as a starting point in introducing evolutionary computation. The problems of chaos, chance, non linear interactivities and temporality being solved by biological species are proved to be in equivalence with classic method of optimization[15] .

Evolutionary computations techniques that contain algorithms based on evolutionary principles are used to search for an optimal or best possible solution for a given problem. In a search algorithm, number of possible solutions is available and the task is to find the best possible solution in a fixed amount of time. Traditional search algorithms randomly search (e.g. random walk) or heuristically search (e.g. gradient descent), explore one solution at a time in the search space to find best possible or optimal solution, which is computationally inefficient as the search space grows in size. Whereas evolutionary algorithms from such traditional algorithms are population based. Evolutionary algorithm performs a directed efficient search by adaptation of successive generations of a larger number of individuals. Genetic Algorithms is one such evolutionary algorithm in finding an optimal or near optimal solution to a problem. In a traditional genetic algorithm, the representation is bit length string. Its approach is to generate a set of random solutions from the existing solutions, so that there is an improvement in the quality of solutions throughout the generations. This implementation is achieved through main GA operators' viz. random selection of two solutions from individuals in the parent generation; performing crossover operation on these two solutions to generate two new child solutions. Crossover operation is performed by exchanging specific elements of the two solutions selected; and mutation operation is conducted on child solutions to further explore the search space for better solutions. Different variations in simple GA approach can be found in literature survey to improve its search capabilities [16] . Representation of solutions of an optimization problem is to be done in a suitable format in GA to deal with reproduction and mutation operators. This format or structure referred as genotype, needs to be easily interpretable to a solution of the problem under study. In a combinatorial optimization problem, representation of a solution in GA is difficult as well as a challenging task. These are problems containing discrete decision variables and are interrelated by logical relationships. As a result, different mathematical models may exist for the same combinatorial optimization problem and this may lead to different representations for the same problem.

As explained above Cheng, Gen and Tsujimura [6] in their paper representation of JSP in GA into direct and indirect type. Further to that, T. F. Abdelmaguid [17] in his paper classified GA representations into Model based and Algorithm based. In our opinion, all representations are algorithm based though they appear to be model based.

In Priority Rule Based (PR) representation, a chromosome is represented as a string of (n − 1) entries (p_1, p_2...p_n) where n − 1 is the number of operations in the problem instance. An entry $p_1$represents a priority rule selected beforehand. Accordingly, a conflict in the ith iteration of Giffler and Thompson algorithm [18] should be resolved using priority rule represented by p_i. It means an operation from the conflict set has to be selected by the p_i ties are broken randomly. In GA domain, a best set of priority rules should be selected. Here simple crossover yields feasible schedules.

In Random Keys Representation (RK) was first proposed by Bean [19] . In this representation, each gene is represented with random numbers generated between 0 and 1. These random numbers in a given chromosome are sorted out and are replaced by integers and now the resulting order is the order of operations in a chromosome. This string is then interpreted into a feasible schedule. Any violation of precedence constraints can be corrected by a correction algorithm incorporated.

In Operation based representation, each gene represents an operation. A chromosome contains as many genes as the number of operations. For example, an nx m JSP there will be nxm genes in the chromosome. Beirwirth proposed a technique "permutation with repetition" [20] which is similar to operation based representation. Fang [21] also proposed a kind operation based representation where string contains nxm chunks which are large enough to hold the largest job number for the nxm JSP. Whereas Beirwirth used a special GOX crossover technique to generate feasible schedule, Fang used a special decoding approach to decode a chromosome into a valid schedule always.

The Preference List based representation (PL) uses a string of operations for each machine instead of a single string for all operations which is a direct representation of processing sequence decision variables. Quite often violation of constraints is encountered which can be overcome by repair algorithm.

In the Machine based representation, [21] the chromosome contains a string of length equal to the number of machines. The sequence of machines in the string is the order by which a machine is treated as a bottleneck machine in the shifting bottleneck algorithm [12] .

In the Job based representation [22] a chromosome is a string of length equal to the number of jobs in the problem under study. Using this representation, a simple algorithm can generate a feasible schedule given sequence of the jobs onto different machines.

4. METHODOLOGY

The reproduction and mutation operators applied to JSP model are generally adopted from Travelling Salesman Problem because of the similarity in representations. Reproduction operators are generally required in GA to conduct the neighborhood search and a mutation operator generally ensures that the solution is not trapped in local minima. The design of both operators is crucial for the success of GA. Among the reproduction operators reported in the literature, PMX (partially matched crossover) [23] , OX (ordered crossover) [24] and uniform crossover [25] are extensively used in JSSP. PMX and OX crossover techniques use either single point or two point crossover. Different mutation operators used are swap mutation, inversion mutation and insertion or shift mutation reported in the literature [17] .

In general, the flow chart for GA can be represented as shown.

Table 1. Results of benchmark instances under different representations.

Problem	Size	No. of Operations	Best Known Solution	OB Best	OB Avg.	JB Best	JB Avg.	MB Best	MB Avg.	PR Best	PR Avg.
mt06	6 × 6	36	55	55	64.889	55	65.712	55	61.822	55	66.648
mt10	10 × 10	100	971	989	1116.02	971	1100.9	992	1145.06	958	1100.42
mt20	5 × 20	100	1206	1220	1394.47	1206	1383.08	1245	1427.24	1242	1426.54
abz05	10 × 10	100	1259	1275	1394.79	1259	1386.12	1287	1409.87	1267	1390.94
abz06	10 × 10	100	971	958	1072.19	971	1075.97	996	1096.13	978	1080.3
abz07	15 × 20	300	742	734	821.16	742	804.892	751	817.937	730	807.128
abz08	15 × 20	300	758	751	833.362	758	825.982	763	838.59	755	826.954
abz09	15 × 20	300	752	784	877.468	752	849.838	773	873.541	764	859.258
car01	5 × 11	55	7038	7038	8747.84	7038	8694.01	7038	8707.83	7038	8782.28
car02	4 × 13	52	7376	7378	8788.38	7376	8738.23	7221	8817	7166	8881.94
car03	5 × 12	60	7725	7590	9219.19	7725	9195.36	7725	9293.86	7725	9272.51
car04	4 × 14	56	8072	8003	9620.16	8072	9452.62	8276	9697.21	8132	9643.3
car05	6 × 10	60	7835	7873	9207.14	7835	9130.26	7862	9251.68	7862	9407
car06	9 × 8	72	8505	8505	10017.7	8505	9886.82	8505	10229.5	8485	9830.33

car07	7 × 7	49	6558	6576	7673.64	6558	7782.76	6627	7751.89	6632	7738.75
car08	8 × 8	64	8407	8407	9436.29	8407	9500.61	8458	9470.57	8366	9470.64
la01	5 × 10	50	666	666	783.616	666	796.901	674	746.506	666	782.789
la02	5 × 10	50	655	665	748.122	655	774.664	660	745.747	667	757.79
la03	5 × 10	50	617	620	688.729	617	687.389	626	690.773	620	699.527
la04	5 × 10	50	607	595	695.259	607	690.822	619	699.926	602	688.268
la05	5 × 10	50	593	593	640.494	593	658.885	593	606.404	593	699.114
la06	5 × 15	75	926	926	1000.79	926	1021.85	926	958.039	926	1075.5
la07	5 × 15	75	890	890	998.253	890	1015.05	893	983.784	890	994.044
la08	5 × 15	75	863	863	981.109	863	985.795	863	959.264	863	995.054
la09	5 × 15	75	951	951	1051.15	951	1084.34	951	988.331	951	1167.81
la10	5 × 15	75	958	958	1017.01	958	1045.73	958	971.19	958	1089.97
la11	5 × 20	100	1222	1222	1308.89	1222	1334.96	1222	1264.12	1222	1389.85
la12	5 × 20	100	1039	1039	1132.34	1039	1157.8	1039	1104.58	1039	1226.58
la13	5 × 20	100	1150	1150	1248.7	1150	1278.51	1150	1191.37	1150	1314.64
la14	5 × 20	100	1292	1292	1320.59	1292	1348.95	1292	1295.72	1292	1388.82
la15	5 × 20	100	1207	1207	1336.66	1207	1352.88	1227	1368.04	1207	1352.41
la16	10 × 10	100	979	982	1083.26	979	1066.88	988	1088.68	987	1071.61
la17	10 × 10	100	797	793	890.389	797	885.073	832	905.275	807	888.012
la18	10 × 10	100	861	861	962.052	861	967.819	885	976.877	883	977.685
la19	10 × 10	100	875	875	970.966	875	972.302	899	983.686	877	976.925

la20	10 × 10	100	936	907	1022.37	936	1040.62	944	1039.52	914	1041.05
la21	10 × 15	150	1105	1098	1252.76	1105	1247.82	1115	1264.74	1111	1281.85
la22	10 × 15	150	972	988	1146.21	972	1125.68	1031	1161.32	990	1133.83
la23	10 × 15	150	1035	1045	1188.52	1035	1168.18	1037	1180.52	1068	1187.63
la24	10 × 15	150	1004	1006	1135.91	1004	1134.02	1029	1155.7	995	1149.69
la25	10 × 15	150	1040	1055	1177.18	1040	1170.37	1036	1175.69	1058	1178.36
la26	10 × 20	200	1269	1279	1457.86	1269	1424.04	1304	1466.35	1310	1446.12
la27	10 × 20	200	1341	1363	1529.94	1341	1500.85	1421	1539.77	1374	1538.38
la28	10 × 20	200	1301	1295	1454.95	1301	1456.94	1334	1463.16	1284	1453.35
la29	10 × 20	200	1274	1302	1441.65	1274	1416.42	1307	1429.08	1270	1425.34
la30	10 × 20	200	1418	1429	1576.32	1418	1554.81	1444	1592.45	1432	1591.74
la31	10 × 30	300	1784	1784	1927.56	1784	1938.51	1785	1934.69	1784	1933.39
la32	10 × 30	300	1850	1850	2019.59	1850	2029.95	1855	2024.57	1853	2031.31
la33	10 × 30	300	1719	1725	1890.07	1719	1873.95	1719	1871.39	1725	1883.41
la34	10 × 30	300	1757	1782	1942.11	1757	1916.99	1801	1941.67	1793	1941.16
la35	10 × 30	300	1890	1905	2090.49	1890	2079.74	1919	2116.75	1906	2097.97
la36	15 × 15	225	1348	1343	1515.99	1348	1492.07	1385	1519.31	1352	1498.9

la37	15 × 15	225	1486	1506	1674.69	1486	1651.36	1548	1698.19	1496	1687.76
la38	15 × 15	225	1319	1307	1455.03	1319	1474.01	1369	1494.87	1299	1486.33
la39	15 × 15	225	1316	1325	1486.54	1316	1479.86	1383	1519.21	1363	1508.05
la40	15 × 15	225	1296	1330	1469.25	1296	1460.38	1360	1485.68	1338	1492.6
orb01	10 × 10	100	1124	1130	1262.1	1124	1277.16	1150	1304.67	1126	1277.02
orb02	10 × 10	100	924	919	1047.74	924	1031.34	949	1045	931	1060.46
orb03	10 × 10	100	1067	1116	1254.65	1067	1232.93	1080	1262.65	1065	1230.21
orb04	10 × 10	100	1028	1055	1176.34	1028	1154.35	1075	1164.81	1053	1160.75
orb05	10 × 10	100	931	945	1075.54	931	1056.76	969	1109.28	931	1064.55
orb06	10 × 10	100	1046	1093	1248.37	1046	1213.47	1116	1272.66	1067	1233.12
orb07	10 × 10	100	419	415	469.22	419	467.724	425	472.424	418	470.205
orb08	10 × 10	100	928	940	1113.01	928	1075.88	969	1135.78	947	1102.29
orb09	10 × 10	100	949	953	1080.08	949	1059.32	958	1066.74	964	1079.28
orb10	10 × 10	100	977	989	1150.48	977	1125.03	991	1128.54	980	1122.7

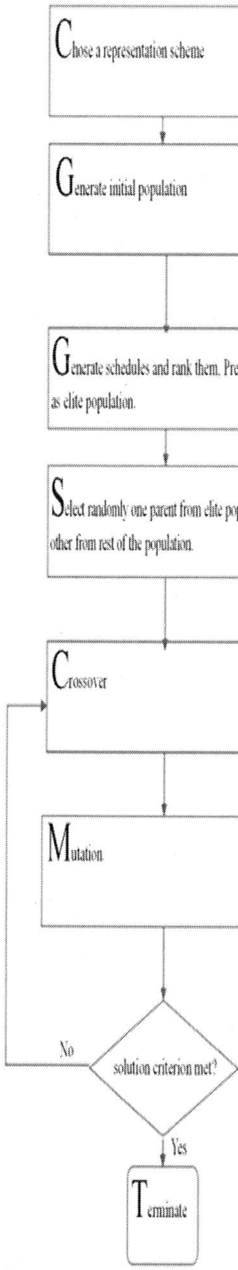

Figure 1. GA-flow chart.

5. RESULTS AND ANALYSIS

In our experiment, four representations are used viz. Operation based (OB), Job based (JB), Machine based (MB), Priority rule based (PR). All experiments are conducted with 50 generations and a population size of 1000. Mutation probability varies with 0.1 to 0.9 values dynamically and elite population size is 20%. Reproduction probability used in our experiment is 0.1 Parents in our experiment are selected from two groups sorted out based on fitness value (i.e. minimum make span). Each parent is selected from these groups probabilistically.

In our experimentation, GA is programmed with different reproduction and mutation operators'. Instead of selecting operators randomly as in [17] , we have built-in reproduction operators and are being used across the representations and the benchmark instances. The benchmark problems used in this paper are taken from OR library [26] available in World Wide Web. All the experiments are conducted with a Pentium-4 dual core processor with clock speed of 2.06 GHz and RAM of 512 Mbs. 68 benchmark instances are taken and in the single run, the best and average values are obtained and compared with lower bound or optimum value of the benchmark instance. Results are shown in Table 1. Different graphs generated are also shown below.

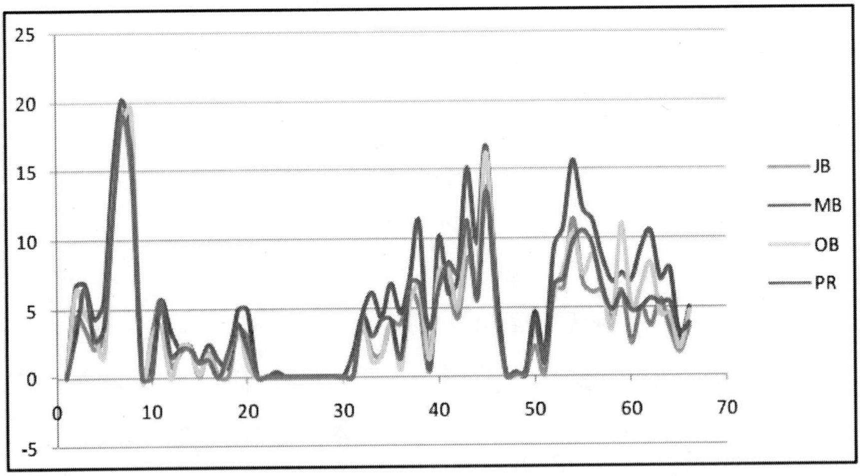

Figure 2. Deviations of different instances under different representations.

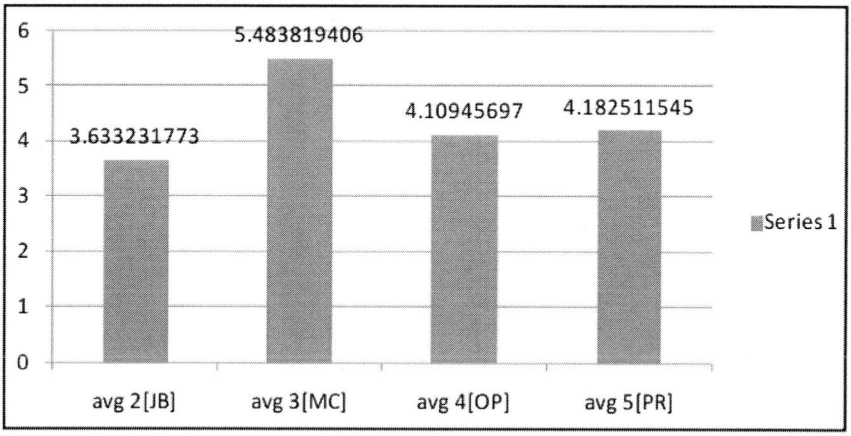

Figure 3. Avg. deviation vs. GA representations.

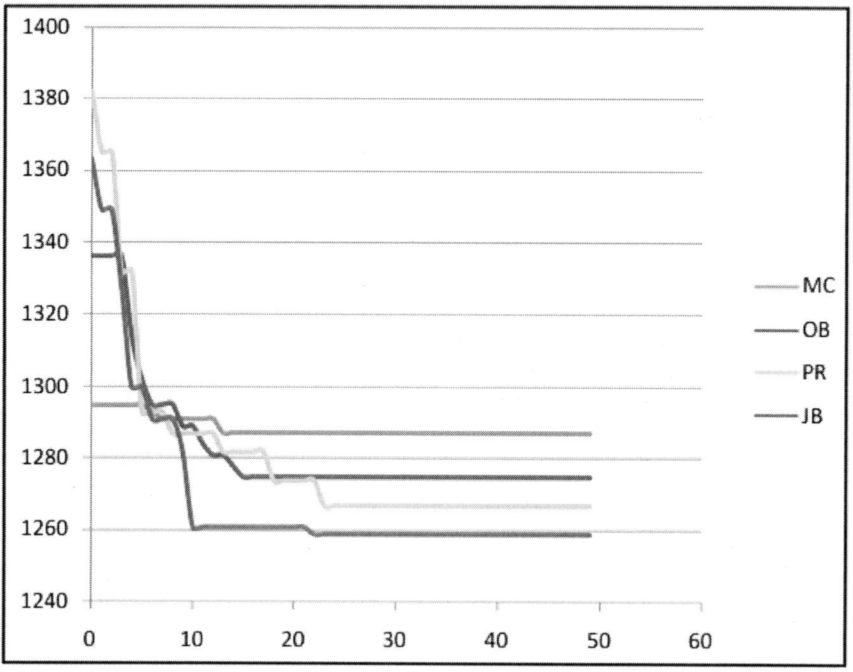

Figure 4. Evolution of ABZ5 under different representation.

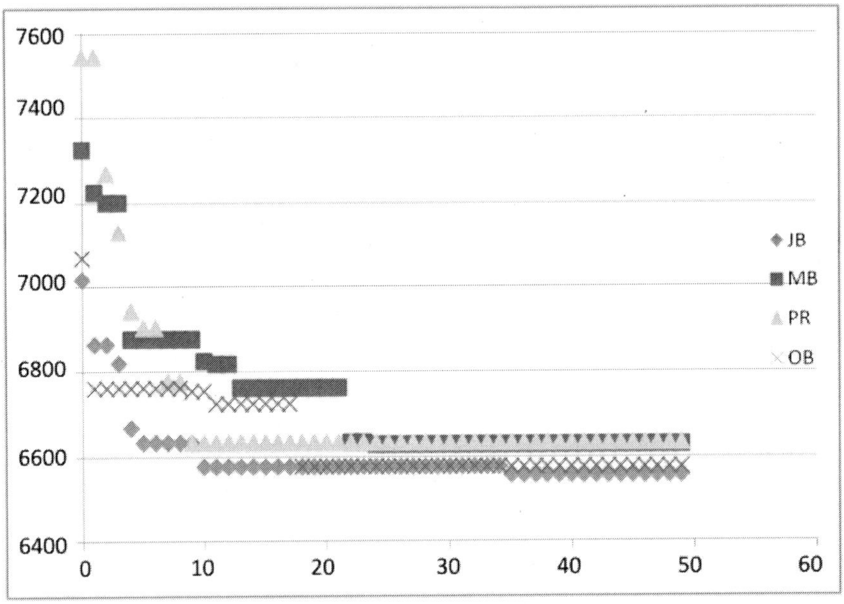

Figure 5. Evolution of CAR-07 under different representations.

6. CONCLUSION & FUTURE SCOPE

Figure 1 shows a plot of % deviations of different instances from the Lower Bound values or Optimum values vs. different representations in GA. It is clear that all representations, across the benchmark instances have shown nearly similar deviations. It is quite clear from the graph that Job Based representations have shown considerable lower peaks. This shows that with the use of proper local search technique it is possible to find the optimal solution.

Figure 2 and Figure 3 show a plot of average deviations of different representations. Except Machine based all other representations have shown the similar deviation. We conclude from this plot that Machine based representation performance is poor and Job based representation performance is better.

The evolution process over 50 generations for the benchmark instance ABZ 5 for instance has been shown in Figure 4 under different representations and convergence of CAR-07 under different representations is also shown in Figure 5. The convergence in case of JB and PR representation is comparatively better than other representations. Whereas JB starts with lesser initial value compared to PR, evolution is faster in case of PR than JB. However, JB could achieve the lowermost value which is why we intend to use this in our further studies. The present work is limited to performance study of different representations of JSP in GA only.

In our further study, we intend to use Job based representation In GA and with the aid of other techniques work to get optimum solutions in possible number of instances.

REFERENCES

1. Brucker, P. (2005) Complex Scheduling. Springer Publications, Berlin.
2. Applegate, D. and Cook, W. (1991) A Computational Study of the Job-Shop Scheduling Problem. ORSA Journal on Computing, 3, 149-156.
3. Balas, E. and Vazacopoulos, A. (1998) Guided Local Search with Shifting Bottleneck for Job-Shop Scheduling. Management Science, 44, 262-275.
4. Jain, A.S. and Meeran, S. (1999) Deterministic Job-Shop Scheduling: Past, Present and Future. European Journal of Operational Research, 113, 390-434.
5. Ponnambalam and Jawahar, S.G. (2007) Hybrid Search Heuristics to Schedule Bottleneck Facility. In: Levner, E., Ed., Manufacturing Systems-Multiprocessor Scheduling: Theory and Applications, Itech Education and Publishing, Vienna, 436.
6. Cheng, R. Gen, M. and Tsujimura, Y. (1996) A Tutorial Survey of Job-Shop Scheduling Problems Using Genetic Algorithms—I. Representation. Computers and Industrial Engineering, 30, 983-997.
7. Anderson, E.J. Glass, C.A. and Potts, C.N. (2003) Local Search in Combinatorial Optimization. Princeton University Press, Princeton.

8. Manne, A.S. (1960) On the Job-Shop Scheduling Problem. Operations Research, 8, 219-223.

9. Roy B. and Sussmann, B. (1964) Les Problemes d' Ordon Ordonnancement Avec Constraints Disjunctives. SEMA, Note D.S., Paris.

10. Abdelmaguid, T.F. (2009) Permutation-Induced Acyclic Networks for the Job Shop Scheduling Problem. Applied Mathematical Modeling, 33, 1560-1572.

11. Carlier J. and Pinson, E. (1989) An Algorithm for Solving the Job-Shop Problem. Management Science, 35, 164-176.

12. Adams, J. Balas, E. and Zawack, D. (1988) The Shifting Bottleneck Procedure for Job Shop Scheduling. Management Science, 34, 391-401.

13. Monch, L., Schabacker, R., Pabst, D., et al. (2007) Genetic Algorithm Based Subproblem Solution Procedures for a Modified Shifting Bottleneck Heuristic for Complex Job Shop. European Journal of Operations Research, 3, 2100- 2118.

14. Bowman, H. (1959) The Schedule-Sequencing Problem. Operations Research, 7, 621-624.

15. Sivanandan S.N. and Deepa, S.N. (2008) ISBN 978-3-540-73189-4, Springer Publications.

16. Yun Y.S. (2007) GA with FUZZY Logic Controller for Preemptive and Non-Preemptive Job Shop Scheduling Problems. Computers and Industrial Engineering, 3, 623-644.

17. Abdelmaguid, T.F. (2010) Representations in Genetic Algorithm for the Job Shop Scheduling Problem: A Computational Study. Journal of Software Engineering and Applications, 3, 1155-1162.

18. Giffler B. and Thompson, G.L. (1960) Algorithms for Solving Production Scheduling Problems. Operations Research, 8, 487-503.

19. Bean, J. (1994) Genetic Algorithms and Random Keys for Sequencing and Optimization. ORSA Journal of Computing, 6, 154-160.

20. Bierwirth C. (1995) A Generalized Permutation Approach to Job Shop Scheduling with Genetic Algorithms. OR Spektrum, 17, 87-92.

21. Anderson, E.J. Glass, C.A. and Potts, C.N. (2003) Local Search in Combinatorial Optimization. Princeton University Press, Princeton.

22. Holsapple, C.W. Jacob, V.S., Pakath, R. and Zaveri, J.S. (1993) Genetics-Based Hybrid Scheduler for Generating Static Schedules in Flexible Manufacturing Contexts. IEEE Transactions on Systems, Man, and Cybernetics, 23, 953-971.

23. Goldberg D. and Lingle, R. (1985) Alleles, Loci and the Traveling Salesman Problem. Proceedings of the 1st International Conference on Genetic Algorithms and Their Applications, Los Angeles, 1985, 154-159.

24. Davis, L. (1985) Applying Adaptive Algorithms to Epistatic Domains. Proceedings of the 9th International Joint Conference on Artificial Intelligence, 1989, 162-164.

25. Syswerda, G. (1989) Uniform Crossover in Genetic Algorithms. Proceedings of the 3rd International Conference on Genetic Algorithms, San Mateo, 2-9.

INDEX